TRAINS AS WEAPONS OF WAR
MONISHA RAJESH

Sitting up, eleven-year-old Simon Gronowski strained to listen in the darkness. With an ear-splitting screech the train had braked and come to a standstill, and he could now hear voices and what sounded like footsteps. Gunshots rang out followed by shouts in German, but with no windows and nothing more than a small hatch in the wall of the cattle truck in which he rode, Simon had no idea what was happening outside.

A few minutes later the train set off again. In the arms of his mother, Chana, Simon soon fell asleep on the straw-covered floor, where they lay surrounded by fifty other people, many of whom were crying and groaning. It wasn't long before another disturbance woke him. The truck's doors were open, a cold wind billowing through, and he saw that two deportees were preparing to leap from the moving train.

Taking him by the hand, Chana drew Simon towards the doorway. This was what he'd been practising for, leaping off bunk beds back at the transit camp. But now, with the ground rushing by, Simon was too nervous to jump. The two others leapt into the moonlit night and Simon perched on the edge of the doorway, his feet dangling. Chana lowered him by the shoulders until he was standing on the footplate between the ground and the truck. Suddenly the train braked and Simon took his chance, landing squarely on the opposite tracks, without a scratch or bruise. Waiting for Chana to follow, he looked back as the train sped up and she called out in Yiddish. 'It's going too fast!' Those were the last words Simon would ever hear from his mother. With a shrill whistle the train steamed on towards its final destination at Auschwitz.

His eyes moist, ninety-one-year-old Simon Gronowski is now sitting behind the desk at his home in Brussels, recounting to me the events of the extraordinary night of 19 April 1943. After jumping, he had stood waiting for Chana in the cold. Through the darkness he heard cracks of gunfire and screaming. Simon quickly realised that if his mother attempted to escape she would only end up in the arms of the Gestapo, so he turned and fled through the woods and across the wheat fields of Flanders.

'My first idea was to go back to my cattle car and get back in with my mother so the Nazis wouldn't find me outside the train,' Simon says. 'If the guards found me they would have shot me immediately. I was all alone. Two or three jumped out before me but I had no idea where they were.'

Pressing a hand to the side of his head, Simon stares in silence at the desk before looking up. 'To get back to my mother I would have had to run past the Germans, so I turned left and I ran.'

Between 1942 and 1944 the Nazis deployed 28 trains to deport the 25,490 Jews and 353 Roma detained at the Dossin military barracks, a transit camp in the city of Mechelen in Flanders, to Auschwitz-Birkenau. Early in the morning on Monday 19 April 1943 – the eve of Passover – a train drew up

THE
UNTOLD
RAILWAY
STORIES

EDITED BY
MONISHA RAJESH

THE
UNTOLD

FEATURING
JACK CURTIS • SHAHNAZ HABIB • CLARE HAMMOND
LEON MCCARRON • ANDREW MARTIN • OMAR MUSA
MARK OVENDEN • YVONNE A. OWUOR • VICKI PIPE
FELICITY SPECTOR • SAM WILLIAMS

RAILWAY

STORIES

DUCKWORTH

First published in the United Kingdom by Duckworth in 2025

Duckworth, an imprint of Duckworth Books Ltd
1 Golden Court, Richmond, TW9 1EU, United Kingdom
www.duckworthbooks.co.uk

A catalogue record for this book is available from the British Library

Book design by Danny Lyle
Typesetting by PDQ

Printed and bound in Great Britain by CPI Ltd, Croydon, CR0 4YY

The authorised representative in the EEA is Easy Access System
Europe, Mustamäe tee 50, 10621 Tallinn, Estonia.

Hardback ISBN: 9780715656082
eISBN: 9781914613951

CONTENTS

TRAINS AS WEAPONS OF WAR by Monisha Rajesh   1

THE LAST TRAIN IN SYRIA by Leon McCarron   15

STATION GARDENS by Andrew Martin   35

THE RAILWAY HOTEL FOR OFFICERS
by Clare Hammond   53

THE FREEDOM RAILWAY by Sam Williams   77

MISADVENTURES IN MAPPING by Mark Ovenden   107

SOMEWHERE AND NOWHERE by Vicki Pipe   127

A TIMETABLE FOR GHOSTS by Yvonne A. Owuor   147

THE BAKU BARNSTORMER by Jack Curtis   173

IRON BRAVERY by Felicity Spector   199

THE NORTH BORNEO RAILWAY by Omar Musa   213

HOW NOT TO HACK YOUR TRAVEL
by Shahnaz Habib   229

About the Authors   244

THE
UNTOLD
RAILWAY
STORIES

Memorial to the Belgian Resistance action against the twentieth convoy, 19 April 1943.

in front of the barracks. For the previous nineteen deportation trains that had travelled from Mechelen to Auschwitz, the Nazis had used third-class passenger trains. On those trains, the doors were locked but the windows remained open and a number of deportees had successfully escaped with assistance in the form of tools and weapons from Resistance fighters: 13 in total from the first 15 transports; up to 248 from transports 16 and 17; and 67 from transports 18 and 19.

According to Dr Laurence Schram, an expert on the genocidal deportation of Jews and Roma from Belgium, the SS in Brussels realised they needed to adapt the deportation procedure to ensure the trains arrived in Auschwitz unhindered. Starting with Simon's train, the convoys would be composed of goods wagons and would leave at night, with a reinforced guard to prevent Jews from jumping off the trains. In accordance with new law, the twentieth convoy comprised thirty cattle trucks with doors secured by barbed wire and barred hatches instead of windows. The trucks blocked the view of residents across the road, rendering them unable to witness the loading of deportees onto the train throughout the day.

Simon and his mother were two of the 1,631 deportees packed onto train 801 who sat in darkness until the 10 p.m. departure. The deportation of Jews was common knowledge among Belgian society and the Resistance against the occupation was strong, but Simon is adamant that their final destination was unknown.

'No one knew. Nobody. They thought they were going to be taken to a labour camp somewhere. The Nazis drew a veil over the fact that it was a death camp we would be sent to. I was there for a month and we didn't hear the word Auschwitz even once.'

A month earlier, Simon, his eighteen-year-old sister Ita and their mother had been rounded up by the Gestapo and taken to the transit camp. At the time of their arrest Simon's father, Léon, had been in hospital recovering from lung problems that he'd developed as a miner and he was now in hiding in Brussels.

In the face of pogroms and antisemitism in Poland, Léon had fled the country in 1921 and entered Belgium illegally as an undocumented migrant. He had married Chana, a Lithuanian, and their two children were classified as foreign nationals until the age of sixteen, when Belgian law dictated that they could choose their nationality. When she turned sixteen, Ita had chosen Belgian nationality, and at the time of the family's arrest the Nazis were not deporting Belgian Jews, so Ita was detained separately from her little brother and mother.

'I said goodbye to my sister and of course I had no idea I would never see her again,' says Simon. 'She was not deportable but she was not free either.'

Other deportees had made plans to escape from the train with members of the Resistance, who had been smuggling in weapons in food parcels up until the day of departure. In the barracks children had been practising their jumps, but it was implausible that gas chambers awaited their arrival.

It was only after the war that Simon would learn what had taken place that night on the twentieth convoy.

With little time to prepare, three young members of the Resistance – old schoolfriends named Youra Livchitz, Robert Maistriau and Jean Franklemon – had devised a plan to ambush the train and free as many deportees as possible. They were aware of rumours about the destination and had chosen a spot east of Brussels, near the town of Boortmeerbeek, knowing that the deportees had a greater chance of making it to freedom within Belgium's borders, where the population was more likely to help. They chose a curve in the tracks where the train would be travelling slowly, one that was close to woods in which the escapees could hide from the Gestapo.

A little after 10 p.m. the trio set off on their bicycles armed with nothing but a single pistol and some pliers. It was a windy night and clouds flitted across the sky as they rode towards the railway line. Robert placed a hurricane lamp wrapped in red paper on the tracks to serve as a makeshift stop signal and the trio went into hiding in nearby bushes and trees. They waited. Minutes passed, then in the distance Robert heard the sound of a whistle. The train was approaching. Slowing into the bend, the engine rolled over the hurricane lamp, then came to a standstill to the tremendous sound of braking. With no time to waste, Robert bolted towards the nearest truck and began to cut open the barbed wire with the pliers, pulling open its doors. Inside, he was greeted by the shocked faces of deportees unsure whether or not to jump. With encouraging shouts in both French and German, he spurred them on, handing each one a 50-franc note to enable them to return to Brussels. This was to be the only successful and documented ambush of a deportation train throughout the years of the Holocaust.

Although Simon didn't escape at the point of ambush, the action fired up a number of those on board who managed to break open doors and saw at bars on the hatches, allowing further escapes. In total 233 prisoners fled the train, of whom 118 survived.

After three nights on the move across Europe the train arrived in Auschwitz. It carried 1,398 deportees – 877 were immediately murdered in four newly enlarged gas chambers, Chana included.

'I ran through the night,' says Simon. 'Because I was a cub scout I was prepared and I knew I would get out of it somehow. Before we left the barracks my mother gave me a 100-franc note which I put in my sock. She was already preparing my escape. While running, I was humming to myself. I hummed "In the Mood" by Glenn Miller. It was my sister's favourite song and she used to play it on piano. She loved jazz...' He smiles at the memory of his sister. Five months after Simon's escape, Ita was also transported to Auschwitz and gassed on arrival.

At dawn Simon came upon a village and chose to seek refuge in a small worker's house – the big houses were often occupied by the Nazis. When a woman answered the door he claimed he'd been playing nearby with children and got lost. This explanation made no sense: a little French-speaking boy in Flanders, fifty miles from Brussels, wearing torn clothes covered in mud. Without asking any questions, the woman asked her neighbour to take him by bike to a kindly gendarme, who offered to help. Simon fell sobbing into his arms and told him everything that had taken place. The gendarme's wife fed and bathed Simon and dressed him in her son's clothes. The gendarme then took him back to the station and helped him to buy a ticket for the train to Brussels.

On arrival Simon went into hiding in a family friend's home, while his father lay low in a separate location. For seventeen months Simon moved between houses, scared that he would be caught and praying that God would bring back his mother and sister.

'I changed houses three times during that period,' says Simon, holding up three fingers on a small hand. 'And every time I arrived I went up to the attic to see if there was an escape over the roofs. To see if I could flee. Every time the bell rang I was scared stiff. During this time I only left the house two or three times in total.'

On 3 September 1944 Brussels was liberated and Simon was reunited with his father.

'We were waiting for my mother and my sister to come back. But when the Allies went into Germany they discovered all the concentration camps and they found the gas chambers and the crematoria and the mountains of bodies, and my father realised that they were never coming back. He died on 9 July 1945 of a broken heart.'

Now a practising lawyer and jazz pianist, Simon sits surrounded by papers, legal memorabilia and books, bringing out a postcard Ita had written to him and thrown from her deportation train. It detailed how 'we are leaving and being taken to work on farms in Holland', the rumour the Nazis had propagated to keep people calm. It had reached Simon's hands long after her death. But in spite of the horrors of his youth, Simon remains positive about the human condition.

'I remain an optimist. I don't want to send out a message of sadness but one of hope and happiness. I say that life is beautiful. I fight for peace and equality among all people.'

Nonetheless Simon has given up his Jewish faith and become atheist, believing that if God had existed he would never have allowed the Holocaust to happen.

'What's beautiful is that witnessing it directly, you then become witnesses in your own terms. Another reason I speak is because I want to thank those who risked their lives to save me. I want to thank the three young men who attacked my train.'

Youra Livchitz went on to be captured and executed in February 1944, refusing to be blindfolded in front of the firing squad. Jean Franklemon was arrested in August 1943 and deported to concentration camps in Germany but survived the war, dying in 1977. Robert Maistriau was arrested and transported to the Buchenwald concentration camp in May 1944, but survived and moved with his family to the Congo before his death in Belgium in 2008.

*

In September 1825, the first steam locomotive carrying passengers on a public line marked the beginning of a new way of life. But it was only a matter of time before this pioneering mode of transport would be harnessed for war. Railways rapidly evolved from transporting freight such as limestone, iron and coal to carrying buoyant tourists to the glorious British seaside. But militaries soon noted the train's ability to transfer populations by the load, and quickly adapted their purpose to transport troops and weapons to the frontline. This allowed battles to go on for much longer and with increased deaths and destruction, owing to the larger supplies that could now be carried to sustain fighting. At the Wannsee Conference, in the Berlin suburbs in January 1942, it was noted that trains would be a key requisite to achieving the Final Solution, and to date they remain a powerful symbol of the genocide of European Jews.

Mass transportation of detainees would soon be replicated by the Khmer Rouge during the Cambodian genocide, which resulted in the deaths of two million people, a quarter of the country's population. In April 1975, Pol Pot recognised the potential of the French-built railways to forcibly relocate Cambodians from big cities to the countryside on an industrial scale, where they were put to work in horrific conditions.

Survivors' testimonies reveal how civilians were separated from family at different stations, packed into carriages and taken off to unknown destinations, where they were offloaded, disorientated and miles from home, sometimes being forced back on the trains and transferred several times further. The journeys were undertaken in gruelling conditions: searing heat; without food or water; many falling sick before they arrived at their destinations, where they were condemned to hard labour, starvation or massacre. Almost completely destroyed by war, Cambodia's railways were neglected until the government pledged a railway revival that has finally reinstated a handful of regular passenger and freight services.

Around the world, railways reveal much about a country and the nature of its people. Whether it's the silent carriages on the Shinkansen in Japan with their neat, single-file boarding, or the raucous free-for-all style of travel in India, trains lay bare cultural practices, showcase cuisine and present themselves as microcosms of the societies through which they run. Modern-day travellers ride these trains, gazing from open windows, taking in the views of coconut groves or neon mega-cities flashing by, many of them oblivious to the train's sinister origins.

In 2010 I spent four months travelling the length and breadth of Indian Railways in awe of its charms, each train exuding a character and personality of its own, illustrated by the regional snacks hawked up the aisles to my companions singing songs in various dialects. A glorious gift from the British, say many an apologist for empire of India's railways, but what is ignored is how the East India Company conceived of the railways as a means to bleed the country of its resources, extracting cotton, iron, coal and tea. By the time of Indian independence in 1947, the British had built 54,000 kilometres of track along a network that rippled out into the nooks and crannies of the country – not a benevolent boon to Indian citizens, but a deliberate effort that allowed the British to establish military control over Indians who not only funded the construction of the railways but weren't initially allowed to travel in the same compartments as their colonial overlords.

It was the same story in Cuba. One of the first countries outside Europe to establish a railway, Cuba's first train arrived as early as 1837. With no interest in Cuba's slave-based society, its wellbeing or economic position, the British built the railway to fill the pockets of the sugar barons who used the trains to transport sugarcane from slave plantations to mills and then from mills to the coast for export.

Indeed, after the Slavery Abolition Act of 1833 the UK government paid £20 million in compensation to former slave owners who invested a large percentage of that fortune into the founding of Britain's railways. One notable example was Charles Lawrence, a Liverpool merchant awarded compensation after the freeing of slaves on the Fairfield estate in St James in Jamaica, which produced sugar and rum. In a database collated by University College London's Centre for the Study of the Legacies of British Slavery, it is document-ed that Lawrence invested £99,450 into British railways, including £60,400 in the Liverpool and Manchester Railway, for which he was chairman. To further add insult to injury, the railways themselves were then used to transport sugar plundered from the West Indies and also cotton, imported from slave plantations in America's Deep South.

*

For all the stories of railways used for harm, there are magnificent moments in history where railways and their workers emerge as heroes. When the atomic bomb struck Hiroshima on 6 August 1945, the city was all but destroyed, a handful of buildings left standing among the smouldering debris. And yet the trains were up and running almost immediately, offering the only means for survivors to flee the fallout. Aged eleven, Tetsushi Yonezawa and his mother had been riding in a Hiroden street car when the bomb struck, but the metal frame withstood the blast, sparing them from injury and death. In desperation they ran through the burning city, amid a downpour of black acid rain, arriving at Yaguchi station where they found three trains parked and more than a thousand people clambering across burnt bodies to get on board. Together, Tetsushi and his mother boarded a train to Shiwaguchi station, which he credited with saving his life and those of countless others. To this day, Japanese trains remain a symbol of the country's strength in the face of adversity.

'And for me,' says ninety-one-year-old Simon, 'the real heroine was my mother, who sat me on this step of the train and sent me towards freedom and my life...'

*

More recently, Ukrainian Railways have become a saviour since Russia's invasion. A few hours after the first missile and drone strikes on cities across Ukraine, railway workers scrambled to launch evacuation trains to the country's western border. More than four million people were evacuated in the first few months despite the trains being a military target. With an estimated 230,000 employees, Ukraine's railways have continued to provide a lifeline, by transporting hot food and providing humanitarian assistance, with a designated Food Train launched in November 2023 that has since provided more than a million meals. It's a story that's expanded upon in Felicity Spector's essay, one of several in this anthology that explore the role of railways in bringing people together.

Laying bare the often malign intentions behind the creation of railways isn't to cast a negative light on their existence but to allow them to sit alongside the many positives and enhance our understanding of their legacy and future benefits, something that Leon McCarron explores in his essay retracing the Berlin–Baghdad line, the aim of which was to expand German military rule in the Middle East. Clare Hammond, too, learns how the Burmese railways – built in the late 1890s from Mandalay towards the Chinese border – were used by the British to justify overthrowing the last Burmese king, and Sam Williams discovers similar plans of domination in his story of the incredible TAZARA. Further into the book, Vicki Pipe questions the legacy of British trains and wonders to what degree this form of public transport is actually still serving the public in the face of delays, cancellations and confusing ticket types. Mark Ovenden becomes a master of railway maps, and Andrew Martin is content to gaze at the scenery in a delightful chapter detailing how the creation of railways lead to the development of glorious gardens – and why you need to peer between the railings at South Tottenham station.

Ultimately train travel is a joy, and there's much of that to be found in Jack Curtis's attempt to take fifteen trains across ten countries on the Baku Barnstormer, and Yvonne A. Owuor's recollections of her journeys as a child. But even when the best-laid plans come awry, as they do in Omar Musa and Shahnaz Habib's chapters, there are discoveries to be made. Both writers explore complex links between families and histories, travel and mindset because, as Habib explains in the final chapter, the best stories are the ones we don't usually tell.

© Marc Sethi

Simon Gronowski.

Workers keep warm by a fire in the warehouses of the Syrian Railway
Establishment during winter. The maintenance yards close to Aleppo were heavily
damaged during the civil war, but there are hopes that the services will restart.

THE LAST TRAIN IN SYRIA

LEON MCCARRON

On a cold afternoon in February, I sat in a small garden in Aleppo. In front of me was a teacup filled with honey-flavoured Tennessee whiskey. A palm tree overhead swayed gently, moved by a wind that still carried winter on its breath. Opposite me was Roubina Tashjian-Mazloumian, widow of the late owner of a hotel called the Baron. One side of the hotel faced us now, faded brickwork dulled further by a grey sky; windows clasped with Ottoman shutters. Roubina raised her glass.

'I'm sorry it's not Irish,' she said, and took a sip.

I saluted her back, then nodded very slightly towards the hotel. It was why I'd come. At almost any other point in the last hundred years, we'd have been toasting one another inside.

The Baron Hotel was once an icon of Aleppo, though it can be hard to imagine it now. From the garden I could see three storeys, five arched windows on each and balconies on the middle floor. A rusted gateway was locked shut under the space where the hotel's name used to be. It shut down in phases during the Syrian civil war, at one point taking in refugees from the fighting. The building avoided major physical harm from the conflict, but earthquake damage in early 2023 added to the terminal lack of clients, and its doors were permanently shut.

'But it *was* luxury,' Roubina told me.

The Baron was built just a decade into the twentieth century, specifically to give diplomats and dignitaries of a certain stripe a suitably sophisticated place to stay.

'Lots of royalty and statesmen,' said Roubina of the guests.

She listed Charles de Gaulle, Kemal Atatürk, Gamal Abdel Nasser and the last Shah of Iran, Reza Pahlavi. King Faisal declared Syria's independence from the balcony of room 215. In these names alone lies much of the modern history of the Middle East. And then there were the Brits. Gertrude Bell, the traveller and colonial diplomat, was a frequent visitor. The army intelligence office T. E. Lawrence had a long-standing contract in room 202 and left an unpaid bar tab, which somehow doesn't seem surprising. It was later framed for posterity. Agatha Christie wrote chapters of *Murder on the Orient Express* from 203 while her husband worked on archaeological sites in the north-east of the country.

'Apparently she used to like people to know who she was,' said Roubina, smiling, swirling her whiskey. Her father-in-law had told her stories of Christie sitting at parties, waiting to be approached. He was never sure if it was pride that held her back or shyness. Roubina paused.

'And of course, they all arrived by train.'

The Baron was only part of the reason I'd come. The real reason was the railway. In the late nineteenth and early twentieth centuries, a rail project to link Europe to the Middle East passed through here and a family of Armenian hoteliers saw an opportunity to create elegant accommodation for those unenthused by crowded *caravanserai* lodging.

The result was the Baron. From the balcony of Agatha Christie's room, said Roubina, the roof of the train station was just visible. What, I wondered, was left of the legacy of that era, when those who shaped the region – or wrote about it – stepped off a steam train and into a leather armchair? It might seem indulgent nostalgia, but perhaps there was something to be learned. At the very least, how did we get from there to here?

*

On 8 December 2024, a lightning offensive spearheaded by Hay'at Tahrir al-Sham (HTS) reached Damascus and ended fifty-four years of the Assad regime. I arrived in Syria around three weeks later. It would probably have been earlier, but I got food poisoning and spent the early days of the revolution in bed in Beirut, which ever since has provided a disappointing answer for Syrians when they ask me, 'Where were *you* when it happened?'

For over a year, I had been researching a book on the Ottoman-era Hejaz Railway. I had conducted scores of inter-views across four countries, walked for days on sections of track in Jordan and Saudi Arabia, and seen Ottoman archives in palatial Istanbul libraries. My first trip to Syria was under regime control in early 2024, during which time reporting was unsurprisingly limited. Then the collapse of the Assad regime changed everything, for Syria, the region, the railway and, consequently, my research.

When I returned to the country in early January 2025, I revisited the stations of the Hejaz, but first I went north to Aleppo. It was from there that most of the world began to contemplate the fall of Assad after the rebels overran the city in late November. When I arrived, the flag of the revolution flew above the crenelations of the citadel and HTS soldiers waved cars through at checkpoints into the city.

The morning after my visit to Roubina, I walked the short distance between the Baron Hotel and the railway station. It had been called Gare de Baghdad – and the street still bore the same name – because of its centrality on the Berlin–Baghdad Railway.

The Berlin–Baghdad line was primarily funded and engineered by the German empire under Kaiser Wilhelm II. Much of its route passed through Ottoman territory, with the aim of extending German economic and military influence in the Middle East. Inevitably, this made the French and the British nervous. In the early years of the twentieth century, as the project overcame seemingly insurmountable technical challenges to cross the Taurus Mountains, the railway became embroiled in the escalations and tensions that would eventually lead to the First World War.

At the heart of the railway's ambition was Aleppo. When the station was built in 1912, it represented an upgrade of the city's role in line with the times. Aleppo, inhabited for 8,000 years, had been a crossroads between Mesopotamia and the Levant from the earliest civilisations of the Bronze Age all the way through until the Ottoman era. It was, and perhaps still is, a critical junction in any imperial vision.

The architecture reflected the station's prestige. Today, the city has grown around so close it is hard to get a sense of scale, but the façade has the hallmarks of both Europe – functional elongation and sturdy brickwork – and the decorative arched windows and carved motifs of the Ottomans.

Inside, in 2025, the ticket hall was empty. Above the ticket booths, a series of Syrian Arab Republic flags – the old ones – were framed and mounted on the wall. These hadn't yet been removed, although surely their time was drawing close. In an adjacent waiting room, three multi-tiered Ottoman chandeliers hung above a marble floor, and a bizarre mural showed a collection of scenes. Ostensibly they must have been intended to display the greatness of Syria through history and depicted tanks firing shells beside a snowy mountain peak, raiders on horseback attacking the medieval citadel, and a benevolent-looking Hafez Al-Assad watching over two modern trains and a large satellite dish. It reminded me of something a Syrian friend had said of the omnipresent pictures of the Assad father and son around the country – that the family was 'as vain as they were evil'. Artistically misguided too.

In a small side room, twelve young HTS recruits were gathered around a wood-burning stove. The temperature the night before had been -5C. These men were tasked with guarding the station as the interim government tried to get a grip on security. But the cold weather had got the better of them and their weapons were stacked by the door, fingers occupied by teacups rather than triggers. With them was Mahmoud Al-Jasim, the manager of Aleppo Station. He was bundled up in a leather jacket and took me by the arm.

'I'll show you whatever you want,' he said. 'But it won't be pretty.'

The railway compound extends far back beyond the original station and is filled with the memories of what was once the Syrian railway network.

'You have to understand what it was like here before,' said Mahmoud. We walked in long grass that had grown through the tracks and weaved our way between train cars.

'On busy days, like a national holiday, we would have 20,000 passengers coming through this station.' Even on regular days, he said, there were twenty to twenty-five trains passing through and thousands of people on the platforms.

One legacy of the Berlin–Baghdad endeavour was the enduring route which came to Aleppo from Istanbul in the north-west, and its continuation eastwards along what became the Syrian–Turkish border. Eastern Turkey, Tehran and Baghdad were all still viable rail destinations long after the fall of the Germans and the Ottomans. By the 1960s, the railway serviced almost every major city in Syria with over 1,500 miles of track. Routes branched east, reaching Raqqa, Hasakeh and Deir Ez-Zor. Two separate lines crossed into Lebanon. In the south was the Hejaz Railway line, a separate imperial endeavour, leading to Jordan and Saudi Arabia in one direction and Palestinian Haifa on the coast in another.

'We were the nerve centre of the country,' said Mahmoud. 'And now look.'

Most of the stationary carriages were from the 1970s and 80s. Some were sleeper cars designed for trips to Istanbul or Tehran, with small cabins of four beds. Others had been decorated in an 'oriental' style, with fake mosaics and wooden blinds. By Mahmoud's reckoning, there were around three hundred passenger cars still in the station. At the time of my visit, there were precisely zero operational routes on which the cars could be used. The previous year there had been a limited passenger service operating on the coast, between the cities of

Tartus and Latakia, but that too had stopped since the end of the regime. Even the freight network had ground to a standstill, with one exception. As we walked a Russian locomotive was being prepared by the platform, and Mahmoud nodded towards it. It would transport grain between Aleppo and Homs.

'That,' said Mahmoud as it rolled past, 'is the last train in Syria.' In a large workshop filled with American, French, Czech and Russian engines, three engineers stood around a fire smoking cigarettes. Between them they had over a hundred years of service with the railway company, and even with the uncertainty of the times they showed up to work. There were still jobs to do, said one, Mohammed Bilal. In fact, they were due to check the tracks connecting the station with the maintenance yards outside the city. Mohammed pulled on a heavy jacket and asked if I wanted to join for a ride.

Five of us crammed into a yellow locomotive designed for railway repairs. The depot was at a place called Jibreen, ten miles away, on the route east to the desert and Deir Ez-Zor. Mohammed apologised for the broken window and I, in turn, pretended not to feel the chill of the wind. He smoothed the collar of his olive-green shirt and lit another cigarette. He was seated at a small control panel and looked down at a few dozen buttons. Ash fell on the dials, obscuring the numbers.

We pulled out of the station, slowly. The track itself was clear of debris, and Mohammed said infrastructure north and south from Aleppo was generally still operational, as well as that on the coast. The lines to the east, as well as the bridges and tunnels, were long since destroyed in the last thirteen years of war.

Eastern Aleppo drifted past the broken windows like a half-snatched nightmare. Rebels had taken this side of the city early in the conflict, and for four years the area had been besieged and relentlessly hit by regime airstrikes, barrel bombs and shelling. When it was retaken by the government in 2016, tens of thousands were dead and millions displaced. Entire neighbourhoods had been razed to the ground. Against a pale winter sky stood the skeletons of one-time apartment blocks, rebar and steel protruding like bones. Pancaked roofs sloped towards the tracks and rubble was everywhere.

Jibreen was more of the same. Perhaps even worse, somehow, as amongst the devastation of eastern Aleppo there were still some small signs of life – clothes drying on a line beside a makeshift home or smoke rising from a barrel, keeping someone temporarily warm. In Jibreen there were only stray dogs, too tired even to chase the train. Mohammed and Mahmoud pointed out where the militia groups that supported Assad had based themselves – Lebanese Hezbollah here, Iranians there. For years during the heaviest fighting, the railway maintenance yards at Jibreen became a military base, and any building still standing wore the marks of shelling and mortar attacks. One area, now cratered in the middle, had been hit by an airstrike just a few days before, as the Israelis exploited the void in governance after the fall of the regime to destroy Syrian military facilities and weapons stores.

We stopped and walked, picking our way through the war junk. It was impossible to believe that this had once been a hive of activity, with a hundred buildings and six hundred workers.

The landscape was flat and opposite us olive and pistachio trees flowed out in neat rows to the horizon. The ground was fertile, said Mahmoud. In better times, people thought of the countryside here for what could be grown under the soil, rather than the horrors that happened atop it. In one warehouse, a handful of men stood underneath the chassis of a locomotive. There was still a small crew of workers, said Mahmoud, maintaining the grain train. But what happened next, for them and Jibreen, was anyone's guess.

On the way back, Mahmoud told me he'd been to South Korea, Sweden and Norway for work. He'd even been offered a job by a Korean company. Why did he stay, I asked. 'It's my destiny to stay,' he said. He had three children, all in their teens. All they could remember was war. He sometimes wondered about the wisdom of his decision to remain. So many people he knew had left.

'But there's something I heard from a German colleague,' he told me. 'It translates to something like "only dead fish swim with the current". I didn't want to leave, like all the others, and be trapped as a stranger forever. So I stayed. It was a hard decision, but I stayed.'

The next day I went to the famous covered markets of Aleppo. Once a labyrinth of narrow, bustling streets eight miles long, they are a UNESCO World Heritage Site, one of six in the country. The markets hark back to a time when Syria was a crossroads of empires; the meeting point of three continents, and Aleppo the fulcrum. But, like so much in the city, most were destroyed and have yet to be rebuilt.

On a previous visit, I got to know a man called Abdul Hay Kadul, who could usually be found sitting in the courtyard of his Aleppian home, wearing a fedora and surrounded by a thousand trinkets and collectables from all over Syria. Around the year 2000, Abdul Hay had turned his home into a small boutique hotel, mostly for his friends. He still rented out rooms and I had slept there in the past, though the services in the Old City were poor after years of war and it was hard to generate enough electricity to keep warm or heat water.

I wanted to sit with Abdul Hay to talk about the railway because he was a great raconteur and lover of Syrian nostalgia and, after my apocalyptic train ride, I hoped to hear more about what it used to be like. And so we sat with cups of tea, and Abdul Hay leant back in his chair.

'The railway was,' he began, 'the way in which people from the Orient and Occident could meet.'

For an hour he talked of his experiences on the train. As a young man he had a teaching job in Raqqa and would ride east from Aleppo.

'I met tourists, and Syrians of every sort,' he said, listing Assyrians, Kurds, Druze and Bedouin, as well as Alawites from the coast.

'It was something international and something local,' he continued. 'You might have a businessman with a suitcase, or a traveller and their backpack, and then a villager with a chicken on their lap, all thrown together.'

On an earlier research trip, a retired train conductor told me something similar.

'Each carriage was its own little world of kindness and courtesy,' he had said. He remembered how 'unique communities were brought together every time the train left the station'.

Abdul Hay dwelt on this idea of community.

'It's necessary to renovate it, of course,' he said of the national rail network.

I had heard this from many people. Mahmoud had said it and an ex-deputy director of the entire network of Syrian Railways told me so too. They spoke primarily of its economic value and its importance for moving freight across the country. But Abdul Hay saw something equally important in the intangible.

'We are a country of many different colours and we are at our best when we mix them,' he said. At one time, trains passing through Aleppo created a space for easy camaraderie, which today is missed more than ever.

'People who used the train in Syria remember it,' said Abdul Hay. 'It's part of our heritage. Look at the people who came through here to stay at the Baron. And look who went on towards the Hejaz. Once, the world came through here and we helped them on their way.'

*

The following week, I was back in Damascus and, with a few days to spare between other reporting, I went to the grand Hejaz Railway Station. It sits downtown, overlooking a busy intersection, designed to make a suitably striking impression on those preparing to travel south. It is still handsome and, unlike the Gare de Baghdad, wears its ornately decorated façade and marble pillared entrance hall without a hint of European influence. Inside, high stained-glass windows splatter colour across an airy waiting room. But even before the war the station had fallen out of practical usage. Railway tracks were ripped out to make way for real-estate development.

The Hejaz Railway was an ambitious infrastructure project ordered by Ottoman Sultan Abdul Hamid II in 1900 at a time when the empire faced existential threats. Abdul Hamid II's contention was that the construction of over 800 miles of track, from Damascus to Medina, would knit swathes of fragile territory back together.

The railway would largely follow the route of the Hajj pilgrimage and reduce a forty-day journey by camel caravan to less than a week. Mecca was the original destination but, in the end, a deal was struck with local Bedouin to terminate the railway at Medina, allowing them to continue their traditional business of escorting pilgrims for the final section.

There were military and economic considerations too. The railway would shuttle troops to reinforce Ottoman defences against the French on the Mediterranean and the British at the Red Sea. The main line was completed in 1908. By 1914, there was a branch connecting Damascus to Haifa, via Daraa in southern Syria.

If the railway line's construction was remarkable – and it was – then so, too, was the rapidity of its decline. By 1917, large sections already lay in ruins. Arab tribes and British soldiers led guerrilla attacks on the railway during both the First World War and the Great Arab Revolt, when the Hashemites rose up against the Ottomans. T. E. Lawrence, when he wasn't skipping out on his bar bill in Aleppo, played a role in this. His immortalisation in the 1962 David Lean film *Lawrence of Arabia* is one reason why the name of the Hejaz Railway is still known in the West.

In the years since then the railway's fortunes have continued to dwindle. The last train departed from what is now Saudi Arabia in 1925. The line to Haifa began to fall into disrepair after 1948, and Syrian–Jordanian cross-border services halted in 2011. The fragmentation of the Hejaz route throughout the last century mirrored that of the region as a whole, as countries and communities became more forcefully disconnected from one another.

In Aleppo, the passenger station at Gare de Baghdad also housed the workshops, but in Damascus the tasks were split. The Hejaz Station was the glitz and glamour while Al Qadem, three miles south, was where the real work was done. At one time over 750 employees were spread throughout a sprawling network of warehouses and factories that cover nearly fifty acres of a Damascus suburb. It was the largest depot in the country.

There I met Yehya Helwani and Mazen Malla – two old-timers who have spent their careers on the railways. Yehya, who wore a delightful broad and twirled moustache, recently retired after fifty-seven years. Mazen still lives on the compound in an Ottoman-era building. He is the third generation of his family to work on the Hejaz and in his wallet he carries a black-and-white photograph of his grandfather driving a train.

Al Qadem, like Gare de Baghdad and Jibreen, was a military zone during the war and closed to employees. One morning, Yehya and Mazen led me on a long walk through the warehouses. Ottoman-era machinery and locomotives were covered in rust and dust. Tanks and ammunition stores had been hidden in bunkers dug on the grounds of the station and now the earth outside was uneven, potholed and littered with spent bullets.

'When I first came back here, I cried a lot,' said Yehya. 'I can't explain what this place means to me. It's part of me. It's part of all Syrians.'

I heard similar sentiments elsewhere.

'The Hejaz is older than our Ministry of Transportation,' said Na'im Al Karazeh, another railroad employee. 'It's part of the heritage of Syria.' People came to see the historic line and to admire the audacity of its construction, he told me. 'Look at what our country used to be like.'

Pride and dignity were two feelings that came up repeat-edly in conversations, particularly in the context of what has been taken away from Syrians. The railway had at one point awed the world.

'When we worked on the railway, we knew we were part of this long, rich history of Syria,' Yehya told me.

All railway employees in Syria used to be given one free ticket each year to visit Europe. Mazen Malla remembers his father travelling to Germany one summer, taking the route via Aleppo to Berlin.

'I always meant to do it, too, but never found the time,' Mazen said. 'I never thought we'd become so cut off from the rest of the world.' Now the idea of travelling to Europe was unimaginable. 'It got to the point where we were dreaming simply of bread to survive.'

*

It is perhaps useful to reflect on the period when these imperial projects began in Syria and to understand their origin. As agent and beneficiary of the industrial revolution, railway technology spread across Europe to North America in the middle of the nineteenth century, where it became an integral part of westward expansion. In the USA the Transcontinental Railroad, completed in 1869, hastened the journey of freight and passengers across the country, as well as the speed with which the Indigenous tribes and land of the Great Plains and the West were irreversibly devastated.

The Transcontinental was the first of several such projects around the world that used the railways to subjugate and exploit. Others, like the Trans-Siberian, were designed to support colonisation and hold together fragile empires. The unfinished Cape to Cairo was an attempt to link British colonies and make more efficient the extraction and exportation of Africa's riches.

Railway development was weaponised to maximise inequality. It facilitated mass displacement, forced labour and land seizures. On both the Berlin–Baghdad and the Hejaz, labourers worked in torturous temperatures of over 50C. Poor diet and unclean water led to outbreaks of typhoid and cholera. The railways also cut through tribal lands and were used to control troublesome populations. In the case of the Berlin–Baghdad, the railway was also used to deport Armenians from Anatolia during the genocide.

But there were aspects of the two projects that sit outside this paradigm too. The Berlin–Baghdad was a partnership between the Germans and the Ottomans that sought to strengthen both empires, rather than inherently weaken one. Aspects of the Ottoman economy benefited greatly, seen perhaps most clearly in Aleppo, where local traders, especially the city's famed textile merchants, were able to use the connection to Europe to prosper.

The Hejaz, meanwhile, was the first railway in the Ottoman territory that received no financing from foreign powers. Instead, a third of the total cost was raised by means of subscriptions and donations from the broader Muslim community. These came from as far away as Singapore, Persia, South Africa and North America. The British tried to undermine donations from subjects in India and Egypt, fearful of the threat the railway posed to their Red Sea hegemony. Money was sent regardless; propaganda was no match for piety. The railway was designated as *waqf*, an Islamic form of endowment, which meant that the Ottomans did not own the lines. They were, instead, community property. This symbol of gratitude to those who funded it, and the railway's connection to one of the pillars of Islam, is perhaps a reason why its legacy has endured in the region, even as the tracks have faded.

*

An irony of my journeys in Syria is that even as I listened to people speak with nostalgia of train rides which would never be repeated, there has been an explosion of rail projects in the Middle East. Or, at least, a flurry of the paperwork that precedes them.

In the Gulf, the Gulf Cooperation Council (GCC) Railway proposes linking Saudi Arabia, UAE, Oman, Qatar, Bahrain and Kuwait by 2028. There are also plans for metro systems and inter-city travel in several GCC capitals, and there is already a high-speed connection between Mecca and Medina, completing what the Hejaz could not. Iran, meanwhile, is part of the International North–South Transport Corridor (INSTC), which will connect India and the Persian Gulf to northern Europe and Russia, over almost 4,500 miles of maritime, road and rail networks.

Meanwhile, in 2023 Turkey announced plans for the Iraq Development Road to connect the port at Al Faw in Iraq with the Turkish border. It promises economic prosperity to both countries via the high-speed transportation of goods from the Gulf to European markets. By its own estimate, the proposal will cut ten days off the Suez Canal route. Around the same time came the announcement of the India–Middle East–Europe Economic Corridor (IMEC). Apart from confirming that all regional infrastructure projects share the same unoriginality in their naming process, the IMEC seems a likely rival to both the Iraq Development Road and the Belt and Road Initiative – China's transcontinental twenty-first century upgrade to the Silk Road.

It is unlikely the projects will manifest as currently proposed. Some may never get off the page. But according to Alberto Rizzi, a visiting fellow at the European Council on Foreign Relations who I spoke to by phone to make sense of these developments, what we are seeing is a fragmentation of the international system. This is nudging it towards multi-polarity, 'wherein emerging economies stand poised to play a significant role'. In an interdependent, interconnected world, the middle powers – countries without a permanent seat on the UN Security Council, but with significant regional influence – can exert more control than ever before and form new and flexible alliances.

Syria has once more become a viable part of these projects. Assad was a major roadblock, said Rizzi. Without the regime, the possibility of integration into the region seems much more likely. How, exactly, remains to be seen. Turkey, a middle power that for years backed sections of Syrian opposition to Assad, is now in prime position. In late December 2024, the Turkish transportation minister Abdulkadir Uraloğlu said he wanted steps taken 'to restore the railway connection [from Istanbul] to Damascus'. Referring specifically to the Hejaz Railway, Uraloğlu said, 'This project is not just about restoring a railway; it is about reconnecting a historical legacy.'

If routes through Aleppo and Damascus are revived, it should probably not be a surprise. Railways adapt. The Trans-Siberian and sections of Cape to Cairo found new life through tourism. The USA's Transcontinental Railroad is now heavily used for freight. After decades of decline, railways are once more at the centre of audacious infrastructure development plans.

Alberto Rizzi sees opportunity.

'Some of the regional actors have realised that a lack of connectivity is a barrier to growth,' he told me.

Even if there's not exactly a political rapprochement, there is an understanding that they must work together or get left behind. It's possible that railways, developed as a tool of imperialism, may have a more positive role to play in the future.

As ever though, there are considerations beyond the economy. In northern Jordan in 2024, I met a farmer who owned land in the area the proposed IMEC would pass through to join Saudi Arabia to Israel.

'We will never, ever allow anything here until we see peace and a homeland for Palestinians,' he said, when I asked him about the viability of such a connection. He is far from being a decision-maker, but his sentiment is no doubt shared by many.

*

A month later, I returned to the Baron Hotel. I sat again with Roubina and we talked some more about the guests and the stories she remembers hearing about them. We wondered again about Agatha Christie and whether she was prideful or just shy. I went the following day to listen to Abdul Hay talk about the time he took the overnight train from Aleppo to Damascus to catch a plane in the 1980s but missed the flight because of a long delay. He didn't mind too much, he said. Such was the nature of that method of transport. We talked about the railway carriage as one of the few remaining spaces that encourage interaction among strangers, especially in a world of increasing isolation.

One evening, I walked alone to the station. It was silent, the city dark in the absence of a functioning electrical grid. It was hard to imagine Syria ever becoming a transportation hub again, but then perhaps that was merely a problem with my thinking.

'Time moves in a cycle,' Abdul Hay had told me as we sat in his courtyard. 'Aleppo has had many turns of that wheel in the 8,000 years we've been living here. Maybe the next spin is a good one.'

Ealing Common station, 1929. The judging of the District Railway garden competition, forerunner of TfL's current In Bloom. The railwayman-cum-gardener looks nervous, so it's pleasing to report that he would win a first prize.

STATION GARDENS
ANDREW MARTIN

At first, the railways in Britain were seen as brash, dirty and dangerous. In fiction, trains tended to kill people. In *Dombey and Son* by Charles Dickens, the railway speculator, Carker, is run down by a 'red eyed' monstrous express. It 'licked up his stream of life with its fiery heat'. In *The Prime Minister* by Anthony Trollope, Lopez is 'knocked into bloody atoms' by a shrieking Scottish express going at 'a thousand miles an hour'. In *Cranford* by Elizabeth Gaskell, Captain Brown's 'foot slipped, and the train came over him in no time'.

Railways were seen as a force against nature, inimical to the peace of the countryside. In his poem of 1844 'On the Projected Kendal and Windermere Railway', Wordsworth asked, 'Is then no nook of English ground secure/ From rash assault?'

This piece is about the railways' tendency to generate gardens, but in *Dombey and Son* gardens symbolise the lost innocence of a pre-railway world. Dickens gave the name Stagg's Gardens to an area of scruffy market gardens just north of London that he'd known as a boy; they were in the line of fire of the London and Birmingham Railway, the first main line to reach the capital, which it did in 1837. In Chapter

6, we are introduced to this little row of scruffy houses, whose residents 'trained scarlet beans, kept fowls and rabbits, erected rotten summerhouses (one was an old boat), dried clothes and smoked pipes'. By Chapter 15, 'There was no such place as Stagg's Gardens. It had vanished from the earth.' In its place, 'palaces now reared their heads and granite columns of gigantic girth opened a vista to the railway world beyond'. To put it another way, they had been replaced by Euston station.

The name of the gardens was a pun, a 'stag' being a railway speculator. The market gardeners couldn't fight the speculators, and nor could the aristocracy, because the technological revolution of the railways had been accompanied by a social one. As Simon Jenkins writes in *Britain's Hundred Best Railway Stations*, 'capital [had] usurped property as the engine of growth'. All the speculators needed to build their lines was an Act of Parliament; not as high a barrier as it sounds, after the franchise reform of 1832. 'Compulsory purchase overrode the once powerful landed interest. Aristocratic estates lived in terror of seeing surveyors on the horizon. A writer to *Fraser's Magazine* told of "young men with theodolites and chains marching about fields; long white sticks with bits of papers attached were carried ruthlessly through fields, gardens and sometimes even though houses."' In 1839, Thomas Arnold, watching a train skirt the grounds of Rugby School, of which he was headmaster, is said to have remarked, 'Feudality has gone for ever.' (He approved of the development, by the way.)

But, as David St John Thomas writes in *The Country Railway*, 'Except where they helped develop suburbs on the edges of the great cities, railways did not urbanize the countryside but became part of it.' Soon, villages all over Britain would be holding bunting-strung carnivals to mark the opening of stations, and the local gentry would be summoned by fanfares to the top tables of the celebratory lunches. Their anxiety had been misplaced; Thomas suggests it had often been manufactured anyway, that they were 'merely after the best compensation terms. They usually stayed on their estates to enjoy both the cash and the trains.'

*

It turned out that railways rather suited the countryside, or vice versa. According to Thomas, 'artists and engravers portrayed the lines lovingly', and writers of *Murray's Handbooks for Travellers* told readers where to stand for the best views of trains crossing viaducts. (Today, the Balcombe Viaduct, which carries the Brighton main line over the High Weald between the North and South Downs, is a popular Instagram spot. One reviewer on Tripadvisor awarded four stars and 'Would have given it five, but there was scaffolding at one end.')

Towards the end of the nineteenth century, railways presented a friendlier face. They were less stressful, both for passengers (since they were safer and more comfortable) and staff: unions were recognised, working hours reduced. And it became apparent that they were not necessarily all-conquering – that in fact they were mortal. The first railway closure is said to have been that of the Newmarket and Chesterford Railway, which ran between Newmarket and Six Mile Bottom from 1846 to 1851.

The Light Railways Act of 1896 co-opted the railways to the side of those underdogs, the small farmers struggling with the consequences of the agricultural depression of the 1880s. The Act allowed the building of country lines on the cheap, for the carriage of crops to market – and passengers, as long as they didn't mind being shunted along with the goods. These lines disturbed the countryside about as much as a small stream would have done. Level crossing gates were not required and signals were few, but then, as the old railway joke has it, one engine has yet to develop the knack of running into itself. Stations might be wooden shacks, or mouldering carriages. A handwritten timetable might be pasted to the wall, but these shouldn't be taken too literally. Summer trains on the Hundred of Manhood and Selsey Tramway, which ran from Chichester to Selsey, would stop if enough passengers wanted to pick lineside flowers. The Derwent Valley Light Railway, which ran south from York to Selby, offered 'Blackberry Specials' in the 1920s.

Writers began to see railways in a mellower light. In *Railways and Culture in Britain: The Epitome of Modernity*, Ian Carter identifies 'Cuckoo Valley Railway', a short story of 1893 by the Cornish folklorist Arthur Quiller-Couch, as the first depiction of a fading railway line within a rural idyll: 'We climbed on board, gave a loud whistle, and jolted off. Far down, on our right, the river shone between the trees, and these trees, encroaching on the track, almost joined their branches above us. Ahead, the moss that grew upon the sleepers gave the line the appearance of a green glade…' As Carter writes, 'Quaintness replaced puissance. Whimsy supplanted power.'

The moss on those sleepers prefigured many railway pastorals. Edward Thomas's poem 'Adlestrop', depicting a country train in a somnolent station, has come to symbolise the lost peace of Edwardian England. As 'No one left and no one came', the poet contemplates the station name on the platform sign:

…And willows, willow herb and grass,
And meadowsweet and haycocks dry,
No whit less still and lonely fair
Than the high cloudlets in the sky.

In *The Railway Children* by Edith Nesbit, the children enter a floral world when they move from London to a cottage near a country branch line. The porter at the local station, Perks, gives the children produce from his garden. They talk to him while reclining on the 'hot' grass of a railway bank.

Here, from John Buchan's novel *The Thirty-Nine Steps*, is Richard Hannay, another exile from the capital, alighting at a Scottish station after his fraught escape from the scene of a murder in which he might be implicated:

About five o'clock the carriage emptied, and I was left alone as I had hoped. I got out at the next station, a little place whose name I scarcely noted, set right in the heart of a bog … An old station master was digging in his garden, and with his spade over his shoulder sauntered to the train, took charge of a parcel and went back to his potatoes. A child of ten received my ticket, and I emerged on a white road that straggled over the brown moor.

The country station was promoted as an idyll by the railway companies and was perhaps regarded as one both by its users and employees. Wages were not high, but the branch line provided 'interest, pride and security', according to David St John Thomas. In charge was the station master, a patrician figure of the locality: people would go to him for references.

He was smartly uniformed but prone to hang his gold-braided coat on a fence post to dig either his own garden (attached to a house adjacent to, or actually on, the station), or the garden of the station itself.

Anyone he interviewed for a job as a porter would be asked their thoughts about the station garden, and that young aspirant probably would *have* thoughts, since he (and it would certainly have been a 'he' until the Second World War) would either have come from the countryside or be keen to integrate himself into it. He might help out with the local harvest, supplement his income by setting traps and snares along the line, and, as David St John Thomas writes, 'if he changed jobs it would be for a local shop or farm rather than another station'.

It was completely logical that gardening should occur. At most, there might be half a dozen trains in each direction a day to ruffle the peace, so there was time for it, and the railway line itself blazed a floral trail. The Railway Regulation Act of 1842 required that the tracks be fenced off (which they are not in France, for example), fortuitously creating a haven for wildflowers – and the ones namechecked by Edward Thomas in 'Adlestrop' were among those commonly seen from trains.

Many of the hundred or so railway companies that existed before the railway 'Grouping' of 1923 that created the 'Big Four' railway companies offered cash prizes to staff for the best-kept station gardens. In an article headlined 'The Prettiest Railway Stations', for the January 1900 edition of the *Railway Magazine*, G. A. Wade wrote that these schemes had been developed most fully by the North Eastern and the Great Western Railways. He mentions early on a GWR stricture (which probably applied to the NER too), that no prize is given 'however pretty from the garden point of view, unless the offices, waiting rooms and lavatories, etc., are kept unimpeachably clean'.

Wade noted that the NER set aside two hundred guineas annually for division among the winners, adding, 'the station master, of course, takes the large share', as he has been 'the directing agency' and 'has in most cases found the money for seeds, plants, etc.' Repeatedly triumphant was the Castle Howard station on the line between York and Scarborough. 'What traveller going from Scarborough to York has not admired the lovely station of the lovely chateau of the Howards?' asks Wade.

*

I certainly have, even though that station was closed to passengers in 1930, along with another half dozen on the Scarborough line so that people might reach the seaside quicker. Today, Castle Howard station is a holiday home (five stars on Tripadvisor) with a lovely garden. Last year, I rode on a steam-hauled special train from York to Scarborough and a woman sitting near me was making notes about the flowers in the passing gardens, both those of the closed-down stations and of the lineside houses that had never been stations. York–Scarborough is a rural line that would be pretty without station gardens, but Wade stressed that all the companies offering prizes pay 'special attention' to the gardens created in more industrial spots and to the 'difficulties the master, the porters, etc., have to contend with in trying to beautify their station'. He cited 'that famous high bank of nasturtiums at Alverthorpe Station [near Wakefield] ... so covered with those flowers that scarcely a leaf was visible for an area of over a hundred square yards.' This, he wrote, shows 'what can be done even when surrounded by coal pits and under skies more dark than fair!'

In an article of June 1900 for the *Railway Magazine*, 'The Aesthetic Aspect of Railways', Fred J. Husband wrote that a recent prize-winner on the GWR – which he archly refers to as the line that 'hies "Westward Ho"' – was in the Black Country (in his oblique style he doesn't actually name it). Here, stone flags were removed from a wide platform to create flower beds and 'a blaze of colour' that stood in 'great contrast to the smoky surroundings, a vast tract of blast furnaces, collieries, huge iron works, and all the things that are inseparable from a region of hardware manufacture'.

'The Horticultural Environments of Railways' appeared in the 1906 *Railway Magazine*, written by F. James, who was possibly a professional horticulturalist. He opens with a severe injunction against fruit-tree planting on embankments owing to 'the attendant expenditure that would be necessary to the inducement of fertility'. But there are 'many banks where hardy species of flowering plants could with advantage be employed to speed the weary traveller on his way'.

James, too, found the most 'creditable' and 'satisfactory' station gardens on the GWR. This is not surprising, since the GWR – its presence in the Black Country and the South Wales coalfield notwithstanding – was the least industrial and most recreational of the pre-Grouping companies (it called itself 'the Holiday Line') and the one with the most imaginative PR department, among whose more fanciful conceptions was the notion of 'the Cornish Rivera'. James mentions that some of the prize-winners had formed 'an arrangement of flowers in letters', as an advertisement of the company's route. This presumably meant spelling out the letters GWR rather than something more involved like 'Change here for Newquay'.

As F. James turns to recommending ornamental trees and shrubs that might nevertheless 'shelter passengers from winds and other adverse climatic influences', his piece takes on the tone of a seedsman's catalogue: 'Aucuba japonica, berberis in variety, the white and yellow broom, cerasus flore pleno (the double white cherry), the green and variegated forms of euonymus, double furze guelder roses...'

The country station, and its garden, was the focus of village or small-town life. Take that job reference given by the station master. The subsequent application would probably be posted at the station. The station clock would count down the hours until the reply came. If that applicant was successful, the journey to and from work would be by train, or perhaps the job would mean leaving the village altogether. All major life events, good and bad, began or ended at the station, and its flowers might sometimes seem more akin to those given for commiseration than celebration.

Speaking of celebrations, the staff would be particularly avid about their watering, mowing and deadheading if a VIP visit was in the offing. Stations that didn't have gardens might be supplied for a one-off splurge by a railway nursery. There was one at York which supplied flowers and the man-made concomitant of flowers, a length of red carpet, to stations expecting a distinguished visitor. In 1941, the York nursery moved to the nearby village of Poppleton, where it now functions as a volunteer-run garden centre, but it is still a railway nursery, in that it supplies hanging baskets to the Embsay and Bolton Abbey Railway, a heritage line.

Then again, people might go to the country station not to catch a train, but just to sit there – the highest of compliments to the station master. Sunshine would bring out the smell of the creosoted sleepers as well as that of the flowers, which might remind the visitor of the possibilities down the line, while the rails in the heat would occasionally make a restless 'crunk' noise as they expanded in the chairs, or brackets, that held them. But the scent of the flowers would counteract this wanderlust, seeming to say, 'What's wrong with here?'

A high point of station garden judging (or perhaps its *reductio ad absurdum*) was reached in the summer of 1930, when the London and North Eastern Railway fastened a garden bench to the front of a locomotive for the judges to sit on, to expedite the inspection of station gardens in the London suburbs. The *Railway Magazine* reported on this belatedly, in November 1948, as a bizarre historical snippet. They named the locomotive, the *RM* always being very punctilious about such things, Holden 2-4-0 No. 7490, adding, 'The engine ran light [without carriages] in most cases but in some out-of-the-way stations a saloon was attached, and of course the speed was limited during such times when the judges preferred to ride outside. We understand that the experiment was not repeated.'

I wonder if they'd thought of using a motor car. The London and North Eastern Railway, and the other three of the 'Big Four' companies created by the 'Grouping', had advertising posters in the 1930s suffixed with the telling slogan 'It's Quicker by Rail'. The threat posed by the internal combustion engine became a crisis for British Railways after the nationalisation of 1948 and so the stage was set for the appearance, in a puff of black smoke, of the railway 'axeman', Dr Beeching.

Beeching was the protégé of the Conservative Minister for Transport Ernest Marples, who was not so much subject to the influence of the road lobby as at the forefront of it. He part-owned a road-building company and had an advanced driving licence. Opening the M1 in 1958, he declared it 'in keeping with the bold scientific age in which we live', which railways, clearly, were not. Dr Beeching, himself a scientist (recruited from ICI), produced his report, *The Reshaping of British Railways*, in 1961. He became chairman of British Railways the following year.

By 'reshaping', Beeching meant closing more than half the stations on the network and about a third of the route mileage. This is not the place to rehash the pros and cons of Beeching, but we might note, of one of his supposedly clinching arguments – a third of route miles carried only one per cent of passengers – that the same could be said of Britain's roads. Beeching did not appreciate the organic connection between those quiet country parts of the network and its main lines.

John Betjeman became one of Beeching's main adversaries. Betjeman's writings on railways followed in the 'moss on the sleepers' tradition – he enjoyed their eccentricity, their peacefulness, their harmonisation with the landscape. But in 1962, he made a TV film called *In View: Men of Steam* about the Great Western Railway, in which he is uncharacteristically ferocious. He inveighs against 'cold-hearted Treasury nominees who will invent arguments against having railways at all – statistical ones – and of course, there are enemies of the railways in the road haulage industry'. Prolonged footage of a station master gardening at Freshford station on the Wessex Main Line (Bristol to Southampton) seems like an attempt to wind up Beeching and Marples, a refutation of the buzz words in their lexicon: 'scientific', 'rational', 'dynamic', 'technological'. In a provocatively languid tone, Betjeman says, 'There's nothing like the peace of a branch line, where the porter has time to dream of the station competition.' (The best-kept station garden competition, that is.)

The Beeching cuts did away with many of the station gardens. Later rationalisations and modernisations got rid of the gardeners, who, as noted earlier, were drawn mainly from the ranks of porters. That was a varied and ill-defined role, but the core task was carrying passengers' bags, which they could do for themselves after the introduction of lightweight trolleys. Whereas there had been 50,000 porters on Britain's railways in 1948, today there are none and the station gardeners are mainly volunteers.

*

We might start with those who wish it still *was* 1948 (in railway terms): the 22,000 men and women who give their free time to Britain's one hundred and fifty or so heritage railways. The phenomenon, unique to Britain, arose from the trauma of the Beeching line closures and the simultaneous withdrawal of steam traction, which was also on the wrong side of Beeching's balance sheet.

The heritage rail movement is a countrified phenomenon; I can only think of a couple that are headquartered in towns. Several of the stations mentioned in those Edwardian *Railway Magazine* articles as having prize-winning gardens were on lines that were closed, only to reopen a while later on heritage lines. The above-mentioned F. James reported in 1906 that among the recent winners of prizes of five pounds given by the GWR were Dunster, now on the West Somerset Railway, and Arley, now on the Severn Valley Railway. Both still have gardens, the latter especially, and it's hard to take exception the SVR's own description of it: 'This glorious rural station has starred in many film and TV productions, including Netflix's *Enola Holmes* and *Oh, Dr Beeching!* on BBC television. It has extensive, beautifully tended gardens and its own café.' When a steam locomotive stands at the platform of a heritage station, it commands all eyes, but what you notice after its departure are the flowers, whether on embankments, in platform tubs, hanging baskets or the station master's buttonhole.

On the national network, the care of country station gardens has usually fallen to the Community Rail Network, which originated in the early 1990s. To mitigate the effects of staff cutbacks on station upkeep, BR had founded a 'Community Unit', which Professor Paul Salveson, a historian of railways and the northern working class, found 'not-very-well-thought-through'. In response, he wrote a report in 1993, 'New Futures for Rural Rail', in which he coined the phrase 'community rail partnership'. This envisaged set-ups whereby volunteers would maintain and promote a station with support from local councils and the train operating company (Community Rail being a phenomenon of the privatised network).

Today, most of the surviving county branches benefit from the involvement of Community Rail volunteers, whether constituted as community rail partnerships or 'Station Friends'. The telltale signs of both are well-kept station gardens. Paul Salveson – who lives in the station house at Kents Bank, on the Furness Line, where he, along with others, tends the garden and runs a gallery and a railway library – commended to me the gardens at a couple of Scottish stations, Pitlochry and Dumfries, and those all along the Settle–Carlisle line.

Those stations are in naturally beautiful settings, as are those on the Wherry Lines connecting Norwich with Great Yarmouth and Lowestoft. I spent several days last summer riding the Lowestoft branch. I recall the boat on the platform at Cantley in which a pyracantha grows, mimicking a sail; also, the butterflies above the wildflowers at Somerleyton station (brimstone and rare wall brown butterflies thrive there, I later discovered). The sudden intimacy of these gardens is pointed up by the contrast with the wide East Anglia horizon rolling by in-between stations, which is dotted with windmills so far distant that they look half-a-centimetre high.

But perhaps a station garden in a beautiful place is in danger of being taken for granted. What of stations of the kind mentioned by G. A. Wade in the *Railway Magazine* of 1900: the ones not so easy to beautify? He found the nasturtiums at Alverthorpe radiant amid the coal mines. In the absence of coal mines, we have ex-mining towns like Hindley in Greater Manchester, whose station is on the Manchester–Southport line. A station on any such coast-bound route ought to be uplifting and Hindley is, thanks to the Friends of Hindley Station and its secretary Sheila Davidson.

Sheila, the daughter of a miner, grew up just outside Hindley in a house that had no garden, but a family friend let her use a patch of theirs, in which she grew marigolds. When she was sixteen her family moved to a house with a garden – 'Something I'd always wanted.' The Hindley 'Friends' were formed in 2006, and the station had not been gardened for a long time when they took over. Four feet below the surface of an embankment they found a large stone name plaque spelling out 'B.R. Hindley', with an image of a corncrake, the insignia of the local aristocracy. The plaque had been created in the 1980s; today, it's surrounded by a pristine lawn. There is also a wildflower corridor and a formal garden.

There are about a dozen regular gardeners, and local adults with learning disabilities are also involved, via the Friends' connection with the gardening charity Thrive. During lockdown, the Friends grew lavender, which they made into a hedge along the platform. Thrive made wax melts from this, which they sell at craft fairs. In 2024, the Hindley train operator, Northern, bought three hundred packs of them to sell at their annual conference, raising £600 for Thrive. I asked Sheila how many awards the garden has won and she said she'd have to think about that. I later received an email listing thirteen, mainly from either the Community Rail Network or the Royal Horticultural Society.

Officially, Sheila is at the station for two hours on Wednesday and Sunday, 'But *really*, it's every day.' I spoke to her in mid-April. 'We've just had hosts and hosts of golden daffodils,' she said, 'and now the tulips are here.' A train driver on the line recently joined the station Friends, 'just to say thanks for the pleasure I get driving through here'. Other stations on the line have gardens, 'But we are the exemplar,' Sheila told me, and she laughed.

Perhaps the best-known station garden competition in Britain is Transport for London's 'In Bloom', the heir to a competition started by the District Railway in 1910. The scheme is run by Ann Gavaghan, who says that two of the aims are to 'create moments of delight and surprise' and to 'connect stations with their local communities'. In this sense, the gardens are a perpetuation of Edwardian station design, as practised on the Piccadilly, Bakerloo and Northern Line (West End branch) by the architect Leslie Green, who, in the absence of familiar landmarks like trees and fence posts, gave each station he worked on a unique and beautiful tile pattern, most of which survive. Another historical precedent that comes to mind is the design of the moquette (a type of durable woollen fabric) used to cover public transport seats in the golden era of that art form, the 1930s, when Frank Pick, the aesthetically minded Chief Executive Officer of London Transport, requested designers like Marion Dorn and Enid Marx to use green in their moquette patterns. Pick considered green to be 'serene' and relaxing for passengers, since it evoked the countryside.

The efforts of In Bloom competitors are sometimes obvious – in the flower displays on the platforms at West Kensington or Morden, say. 'But not all the gardens are designed to be seen by the public,' says Ann. Last year's overall winner ('Best in Show') was a garden created on what was formerly a stretch of tarmac at Upminster Depot on the District Line, a place normally off-limits to the public. In that case, the 'delight and surprise' came from the bonding between staff who worked on the garden in their spare time. (It also strikes me that a garden in such an unlikely spot is a modern equivalent of that bank of nasturtiums at Alverthorpe.)

Another prize-winning garden, with accreditations from TfL and other bodies, is at South Tottenham station on the Suffragette Line of the London Overground. Here, two staff members, Azaz Khan and Ray McDonagh, have created a garden on what was formerly a rubble slope just beyond the ramp leading to the platforms. It stands on a terrace made from wooden crates donated by the bakery next door. The garden is screened by mandatory railings, but carefully placed mirrors allow the public to see it, which they can also do by peering between the thick railings. In summer, they can also smell its strawberries, roses or lavender, and hear its burbling water feature. Passengers are certainly aware of the garden. On my visit, a man exiting the station spoke to Ray though the railings: 'You're doing a great job! Keep it up!' Ray told me that the whole 'tone' of conversations between staff and public at the station has changed since the garden was created four years ago. There are garden open days and Ray told me that one of the terraces is to become a sensory garden for disabled visitors.

Another former In Bloom winner is a garden on the roof of the building that houses Goodge Street Tube station, whose ground floor was designed by the above-mentioned Leslie Green and to which five stories have since been added. The garden was initiated by a staff member who has since moved to another station. He got it underway by carrying suitcases full of compost up to the roof. Marigolds, nasturtiums, wisteria and wildflowers were grown, alongside tomatoes, beetroot, lettuce, cucumbers and apples, which were shared among the staff. A garden survives on the roof, albeit on a slightly reduced scale.

Looking up at the Goodge Street roof one recent rainy evening, I thought I could see a couple of fronds waving amid the chimney pots – or perhaps not. It didn't matter. The garden was doing its work of absorbing carbon dioxide (of which there's a lot on Tottenham Court Road), in virtuous partnership with the environmentally friendly trains running six storeys below. In the case of the line with moss on the sleepers, power had given way to whimsy, but there is nothing whimsical about the station gardens of today.

The colonial-era Gokteik viaduct is a vital link between northern Shan State and Myanmar's central plains. Now more than a century old, it's in danger of collapsing.

THE RAILWAY HOTEL FOR OFFICERS
CLARE HAMMOND

Navigating the crowds in downtown Mandalay, Khin strode ahead. Sweating in jeans and a long-sleeved shirt, I stumbled after her, catching an occasional glimpse of her shiny dark hair. Then, quite suddenly, she disappeared. Casting around, I eventually spotted the entrance to a compound that I never would have noticed if I was alone.

Slipping through the gate, I found Khin standing in a large and beautifully tended garden, where freshly washed clothing had been spread out on the grass to dry. It was enchanting to step from a crowded street, marked by Mandalay's imposing railway station and the grimy hotels that served it, into this quiet, verdant place. Catching my breath, I took in the old wooden house and the koko trees that shaded it from the late autumn sun. There were two pools beside the house, decorated with mosaic tiles. An elderly woman appeared, raising her hand in greeting as she stepped carefully onto the grass.

'This,' said Khin in a hushed voice, 'is the railway hotel for officers.'

The woman waved us over and signalled that we should sit with her beneath the trees. She was seventy-three years old,

she said, and had lived here all her life. When she was young –
when Myanmar was called Burma and was ruled by the British
– her father had worked for the family of a railway officer. Now,
she had taken over his duties and took care of the house. As
she began to describe her work, we were interrupted by the
sound of a door sliding open. A dishevelled middle-aged man
stepped into the garden, wearing a dirty white shirt and *paso*, a
traditional wraparound sarong. We looked up at him in silence
as he wandered towards us, brandishing a slim cheroot.

'You,' he said, reaching down and taking hold of my hand.

This was William, the chief caretaker of the house and
its grounds. He also held the keys to the building that Khin
and I had come to visit: the last remaining railway library
in Myanmar. When the country was under British rule,
libraries had been opened in the other railway towns that had
once formed the heart of the British state in Burma. But the
others had since closed under the rule of Myanmar's armed
forces, which had seized power in 1962, just over a decade
after independence. The military had ruled the country with
brutal force for more than half a century, only giving up direct
power a few months before my visit, in April 2016, to a civilian
government led by democracy icon and Nobel laureate Aung
San Suu Kyi.

This was a unique period of hope and freedom in
Myanmar. Amid sweeping political and economic reforms,
thousands of political prisoners had been freed. Men and
women who had once been tortured by the regime were
now sitting in parliament. Censorship laws had been lifted.
As travel restrictions were rolled back and the country was
connected to the internet for the first time, an independent
media was thriving.

Khin was my colleague at the *Myanmar Times*, a daily newspaper. She had suggested over lunch one day that we visit this library, because I was researching the history of a railway that for me had come to symbolise the British occupation of Burma. Built in the late 1890s from Mandalay towards the Chinese border, the railway had been used by the British to justify overthrowing the last Burmese king. A powerful commercial lobby had persuaded the Indian Viceroy that a railway was needed to counter French influence in the region, and to open the wealth of China's landlocked western provinces to British trade. There were British sources about these events, but there were significant gaps in them. I wanted to understand the history from a Burmese perspective. How had it been recorded? And how was it remembered now?

'Hello, William,' I said, slipping my hand free from his. 'We'd like to visit the railway library.'

'No,' William replied, gazing at the sky. Then he turned and smiled at me and slapped me on the arm. 'This is a house for officers,' he said, pointing back to the open doorway and the gloomy rooms beyond.

He suggested that we look around, so we all stood up and trooped inside. Built during the colonial period, which began in 1824 with the first of three wars and ended in 1948, the house had hardly been touched since. The reception room was decorated in floral wallpaper and fitted with delicate cream curtains. Lace throws adorned a small sofa. Roses were arranged on an ornate wooden stand. There was an old television, but it was concealed by a lilac satin case, as if to hide its modernity.

William pushed us up a creaking set of wooden stairs. He said he knew nothing much about the history of this house or the railway that extended beyond it. He only knew it was now used as a guesthouse by railway officers. This mostly meant retired army men; the railway department was a home for veterans.

Myanmar was experiencing a time of relative peace, but the country's rail network still served the armed forces, shuttling troops and supplies to the borderlands. There, the military waged a perpetual war on various ethnic groups, including the Kachin, Shan and Karen, who had been promised autonomy at independence from Britain – and who still sought greater freedoms. Many senior railway staff were former soldiers, because operating the lines for this purpose required military expertise.

Upstairs, the bedrooms were small, fitted with threadbare red carpets and four-poster beds. The toiletries provided for the officers were as oddly British and outdated as the decor. Arranged on mirrored wooden dressing tables were safety razors, tubes of Gillette shaving cream, glass bottles of Yardley Eau de Toilette and jars filled with cotton buds.

It was as if, one day, the men who served the colonial administration had packed up and the officers serving Myanmar's regime had moved in to continue their work as if nothing had changed.

Back downstairs, we eventually persuaded William to take us to the library, a low brick building on the other side of the garden. At its entrance, Khin stopped to miaow at a frightened-looking cat and then ushered me inside.

As my eyes adjusted to the light, I saw the central room was dominated by a heavily varnished teak dining table. It was surrounded by teak chairs and overlooked by a painting of Thakin Kodaw Hmaing, a Burmese independence hero and nationalist writer.

'He was the one who wanted to get freedom with his pen,' Khin whispered, gazing up at his severe expression and impressive white moustache.

The library was named after Shwe U Daung, who was also known for his anti-colonial writing. Both men had struggled for freedom from the British, but at what cost? The patron of this library and the man who venerated both of these writers, Win Sein, was no freedom advocate. Instead, he epitomised just how far Myanmar's military rulers had strayed from the ideals of the country's independence heroes. A military hardliner, Win Sein had been railway minister in the 1990s. He was best known for his alleged corruption – and for encouraging military supporters to kill Aung San Suu Kyi. In photographs adorning the library's walls, he was pictured dressed in military uniform and handing out gifts to library staff, including cash.

More famous than Win Sein's support for the library was the funding he provided to one of Myanmar's highest-ever-grossing films, which offers a window into the military's worldview. *Never Shall We Be Enslaved* is a historical war drama covering the period of history I was now researching: the British invasion of Mandalay in 1885, which marked the end of Burmese independence. Its heroes are renegade Burmese generals who ignore their king's demands to surrender to foreign rule.

The film's message is clear: Myanmar's armed forces are the only true patriots, ready to sacrifice everything to defend the nation's sovereignty, while other contenders for power are not to be trusted. It was this view that had empowered the generals to crush the democracy movement and brutalise millions of civilians in the borderlands.

The library itself was a disappointment. It wasn't a library of books about railways, as I had hoped, but simply a library of books that were available to railway staff. The librarian was a surly old man who told me he had nothing that would be useful for my research. Whether or not that was true was hardly important. The books were all wrapped in brown paper and bound with string, and I had no way of knowing what they contained without his help.

As we left, Khin suggested that I try visiting Mandalay's largest independent library and newspaper archive. If a Burmese history of the railway existed, she said, it would be there.

*

The Ludu Library, built by the family of deceased Burmese writer Ludu U Hla, wasn't far from the guesthouse where I was staying and I walked there the next day.

It was housed in an elegant four-storey building that looked from the outside like a modern church, with a red gate leading onto a small lawn and wind chimes spinning in the open doorway.

Inside, the shelves were crammed with the largest collection of books in Mandalay, including thousands of leather-bound volumes that had survived the decades of military rule. Everywhere on the floor lay stacks of old newspapers and magazines that the Ludu family had archived since the Second World War. Newer copies of magazines, including *Time* and *The Economist*, were piled on heaps of propaganda leaflets, periodicals and posters. Everything was sun-bleached the same dusty shade of brown.

But the only relevant book the librarian could find was an official history produced by the socialist regime in 1977 to mark the centenary of Myanmar's first railway. It contained lists of dates and photographs of railway infrastructure that had been blown up during the Second World War, but little else.

The owner of a publishing house happened to be visiting the library at the same time as me and he explained just how difficult it had been to publish independently under junta rule. A tall man with a shock of white hair, he told me that everything had to be submitted to the censors in Yangon and they would tear pages from offending books, only sending them back months later. That wasn't the worst of it. He had been to prison twice, he said, for a total of seven years, for his views and opposition to the military. He knew one writer, though, who was an expert on Mandalay's history, and who might be able to help me. He scribbled down the man's phone number, and before I left the city I went to visit him at his home.

There was nothing much to distinguish H—'s house except for his security gate, which was threaded with plastic roses. The celebrated writer welcomed me into a narrow reception room, where he explained that he wasn't really a writer at all.

'I don't know who I am,' he said, with a smile. That was how it was to be a writer under military rule.

He had been trained in astrology, he said, but he used it only to 'raise the morale of depressed people'. He was also a librarian, and an encyclopaedia. He was interested in the small, logistical details of his country's history, telling me that two boats had once been able to pass, side by side, down the Mandalay canal.

He was also interested in railways, and he took my questions seriously.

'The first history is in oblivion,' he replied, when I asked him about the British invasion of Mandalay, how the palace in its walled city had been sacked and the last Burmese king sent into exile. Drunken British soldiers had set fire to the royal library, turning the meticulous records of an entire civilisation to dust.

Soon afterwards, the British built a railway through the royal city, which they renamed Fort Dufferin, after the Indian Viceroy who had waged war on the Burmese king without a mandate from the British parliament.

'The Burmese people thought trains were evil,' H— told me. 'So England on purpose built a track through the palace, and they put a prison inside too.'

I had visited the palace earlier that week. It was now a tourist site, but also a military base and access to much of the compound was heavily restricted. Large red signs explained which areas were off-limits to 'foreigners', which mostly meant tourists, but also anyone foreign who, like me, was conducting tourist-like activities.

In case the message wasn't clear, an enormous red banner drove the point home, instructing Myanmar's people to cooperate with the military to 'crush' anyone harming national unity.

At the entrance kiosk I handed over my passport. It was required 'for security', the soldier on duty told me, as if I might otherwise be tempted into some transgressive act.

Even so, he seemed relaxed. The paranoia about foreigners that had gripped Myanmar's armed forces after independence from Britain appeared to have eased. It even felt possible to smile at the propaganda signs. But still the presence of soldiers here was a reminder that the military retained significant power, controlling three key ministries, including home affairs, and veto power over amendments to the constitution it had written.

Inside the gates, an old woman offered to rent me her bicycle and I cycled along a tree-lined boulevard that led directly to the palace, alongside the old railway that had been paved over. Peering along roads barred by men with guns, I caught glimpses of what lay beyond: soldiers relaxing in a teashop, a training field with bullseyes for target practice, five men employed with brooms in sweeping a perfect lawn.

At the palace there were a few tourists, but not many. Most of them had congregated around the entrance to listen to their guide talking about nothing in particular. There was very little to say. This wasn't the original palace. It was a replica built in the 1990s, during the peak era of regime propaganda. Its reconstruction was part of a wider campaign to demonstrate the military's pedigree, connecting the generals in the minds of Myanmar's people to the Burmese warrior kings.

But as I looked out from the royal watchtower, it seemed to me that the Mandalay Palace complex, with its training grounds and barracks and officers' houses, resembled Fort Dufferin more closely in spirit than the walled city that had been here before it. Here, in a place where the Burmese people might have remembered the violence of British rule, what remained was a fortress, and a museum that spoke most clearly of an enforced silence. After independence and a brief period of Burmese civilian rule, the generals had started to erase the truth about the British occupation from official histories, replacing facts with propaganda denouncing colonialists and, more recently, neo-colonialists.

They had legitimate reasons to do this; the British occupation had been a deeply traumatic experience. But there was another, more sinister incentive. At the same time as creating an information vacuum, Myanmar's generals had adopted and expanded the colonial administration's ruthless methods of control: its repressive laws, its surveillance networks, its prison infrastructure, its enclaves of extraction. Erasing the historical record had helped them to take from the past with impunity what they needed to hold on to power.

*

Back at H—'s house, I sat absorbing his stories; his head was filled with railway history.

'Another point,' he would say, before launching into a tale of one long-vanished industrial railway or another. He recounted stories of conspiracy and sabotage by Indian and British railway tycoons who competed for business, and the British race against the French to China. It was a colourful history of competition and underhand tactics that touched only lightly on politics and war: the sort of history that had survived the decades of censorship.

Now that Aung San Suu Kyi was in power and there were new freedoms, the boundaries of permitted storytelling were expanding, H— said. The problem was that so much information was still classified. To learn the truth about the British invasion and occupation of Mandalay, and the history of the railway that was caught up in these seismic events, he said, I would need to find a 'well-informed and willing' military officer. Nobody else knew anything about history.

'Even the History department is not interested in history,' he said. The regime had decimated the education system as a means to hold on to power. I suggested that such an officer might be hard to find, and he agreed.

'Most of the generals are illiterate,' he snapped, and then he composed himself.

'Do you know why they removed the Mandalay Circle Line tracks?' he asked, referring to a railway that had once run through the city.

I didn't.

'To make G3 guns,' he said, and he smiled.

*

Mandalay station concourse was cavernous and dark, and it was impossible to avoid the betel sludge that had congealed into a carpet over the cracked concrete floor. It was 4 a.m. – too early for me – but I had little choice. A single daily service ran from Mandalay into the Shan mountains and it left before dawn.

This was a legacy of the conflict years, a time in recent memory when ambushes by ethnic armed groups had been common. It was safe enough now to travel across the Mandalay plains in darkness, but the railway department wanted to maximise daylight hours on the long, tortuous journey towards the Chinese border. In the shadows families slept, wrapped in blankets. Standing among them was a soldier, who stepped out into the dim glow of an industrial light.

'Good day!' he called out cheerfully to me. 'Good day!'

The platforms beyond were divided by high wire fences and it took me some time to find the ticket office. Walking up broken escalators into the depths of the building, I eventually spotted a queue of tourists. Identifiable by their backpacks and guidebooks, they, like me, appeared sleepy and disorientated. One woman who had clearly been walking around for some time was yelling over the noise of the engines.

'But where is platform four!' I heard her shout, as she hurried past me, raising her hands in despair.

Following her to the platform, I climbed into a dark carriage, where I found my seat by torchlight. Folding myself onto a narrow bench, I rested my feet on a sack of leaves and fell quickly asleep.

When I woke, it was growing light. The carriage, with its warm-green walls and slatted wooden benches, had a cosy atmosphere. Young men dozed in one another's arms and couples murmured, their voices rising over the clatter of the train against the tracks. From one end of the carriage came the tinny, crackling sound of sutras playing through a radio. An old woman chanted gently along. A cool breeze drifted through the open windows. We had begun the long climb from the central plains into the mountains.

*

By the time work on this railway began in the late 1890s, it was widely known that the wealth of China's western provinces was a myth. Rather than profiting from China's riches as they had hoped, the British had instead spent vast sums of money trying to crush the nationalist insurgencies that broke out everywhere in the years after the Mandalay war. Heavily outnumbered, in a vast, unmapped and hostile landscape of malarial swamps and jungle-clad mountains, they flooded the territory with Indian Army soldiers and military police. Villages were systematically razed and there were mass executions, arrests, trials and hangings, in a campaign known euphemistically as the 'pacification' of Burma.

But it wasn't enough; to gain the upper hand, the British needed railways. A new central terminus was built at Mandalay and tens of thousands of labourers were employed to hack through the jungles under armed guard. The new lines opened in sections so that soldiers could be shuttled to an advancing front line, gradually expanding British authority between the towns.

By the time the railway to China that I was now travelling on was eventually approved, an uneasy peace had been established. A private company took over the railway system and built the new line, but the terrain was more challenging than anyone had anticipated and costs quickly piled up.

Extracting myself from my companions, I walked to the end of the carriage to stand beside an open doorway. Between the rocks and the ragged trees, as the sun rose, I caught a clear view of the mountains and realised with a thrill that we were already among them. We had been inching along a series of switchbacks and after the last one the sun burned through the mist. Delicate indigo wildflowers growing alongside the track soon fell away to reveal miles of fields filled with sunflowers. Among them were clusters of wooden huts, where families were preparing for the day, lifting rattan shutters and propping them open on bamboo poles.

We were nearing the outskirts of Pyin Oo Lwin, a hill station formerly known as Maymyo. This was once the summer capital of British Burma, where colonial officials and their families retreated from sweltering, tropical Rangoon. Under British rule, the town had been a playground for the elite, a place where officials could escape the heat, but also some of the bleak realities of the occupation. I hoped that in this small town oral histories might have survived and I would find a clearer view of this railway's history than I had in the transient city of Mandalay. Sunlight burst into the carriage, as the fields gave way to long straight roads that led to military compounds, and then we pulled into the dense sprawl of downtown Pyin Oo Lwin.

From the station, I walked to a hotel called Maymyo Sakantha, which I had heard was owned by a local historian. At the reception desk, I asked where the owner was. He was out, the receptionist told me. I asked when he would be back. 'Tomorrow,' she said, looking uncomfortable.

Fortunately, his daughter walked in at this moment. 'You are late,' she said. 'My father died six months ago.'

She offered to take me to visit another man, an Anglican pastor, and drove me through the city's leafy streets to a red-brick church. We walked together through the grounds to the pastor's house, where she left me with a young girl, who showed me into a pale blue reception room with a fireplace and a chest filled with books. It could have been a room in England, except for the old-fashioned Burmese wooden chairs with their finely woven wicker backs and seats.

The pastor was a shy man, who said he didn't know much about colonial history. But he offered to take me to see a friend of his, who lived in a nearby cottage.

'What can you do for me?' said Mr Gabriel, when we arrived. 'That's a joke,' he added. He had white hair, as did his wife, and sparkling eyes.

I told him why I had come and he paused.

'Nobody is recording or remembering,' he said at last, taking care over his words. 'They are just living moment to moment.'

He recommended an official history of the town. When I suggested that it might not contain what I was looking for, both men looked uncomfortable.

'You cannot say that,' said Mr Gabriel. Then he apologised and told me that he had responded from habit.

It was unfortunate, he explained, the result of a life spent looking over his shoulder.

'We have been living in a ditch,' he said, sadly.

But he still looked uncomfortable and a little nervous, as if he feared I would ask too much of him. Yes, this was a railway town and undoubtedly oral histories had survived here. But it was also a military town and therefore a place where stories could not be openly shared. Just a few minutes' walk from the church was the Defence Services Academy, the military's elite officer training school, and many of Myanmar's generals had second homes here. There were cadets all over town, dressed in navy blazers with their hair neatly trimmed into crew cuts, and soldiers on the streets, too, armed with guns. There was a furtiveness in the way these men spoke that I hadn't encountered in Mandalay, a city home to more than a million people where anonymity was easier to achieve. For the rest of my visit we limited ourselves to conversation about our families and the places we had lived.

*

The next morning after a teashop breakfast, I stopped in at the local library, where I found a copy of the official history of Pyin Oo Lwin. But it wasn't filled with propaganda, as I had expected. Instead, it plagiarised heavily and seemingly at random from old British texts. This was strange, but at first I dismissed it as an oversight. The generals might have been obsessed with controlling the narrative about Myanmar's past, but they were far from all-powerful.

The librarian didn't know of any historians in the town, so I walked back to the railway station, marvelling at the beauty of the pines and bursts of pink oleander cast against a deep blue sky. Beyond a ramshackle market, I followed the railway tracks to the station yard, where I found the station master, a handsome man with betel-stained lips who seemed happy to talk.

One of his engineers had a grandfather who had moved from India to operate the railways under British rule, he said. But he was dead now. Everyone in the older generations had died or moved away.

'Most people who built the railway were foreigners,' he said, 'and many of them returned to their places.'

But the foreigners were coming back now that travel restrictions had been rolled back, he said, particularly during winter. They came here to see the old British railway and the town's colonial architecture. To him, there was little to distinguish these two groups of people – the foreigners who had once come to occupy the country and those who now came to look at what was left behind.

My own interest in colonial history suddenly felt like a peculiarly British obsession. I wasn't the only one who had come here by train, admiring the old railway with its switchbacks and mountain views, and asking questions about the past that had little relevance to anyone here. All the tourists who had travelled with me had been doing the same thing. We were like a horde of colonial ghosts, doomed to seeing the world through the lens of the empire – and perhaps because of this, failing to understand anything of what we saw.

The station master suggested that I visit the Governor's House. Once the summer residence of the Governor of British Burma, it's main buildings had been restored and it was now a hotel and also functioned as a small museum. It was owned, he said, by Tay Za – a notorious military arms dealer, whose Htoo Group of Companies was one of the country's largest conglomerates. A handful of well-connected men had become rich under military rule, building business empires that controlled everything from banking and airlines to luxury hotels.

Intrigued, I left the station and walked on through the town, past the sprawling Defence Services Academy compound. From the main road, I turned up a long, tree-lined drive to the Governor's House. A grand teak mansion set in manicured gardens, the house was surrounded by bamboo and ferns. Two vintage British cars were parked in the driveway. There was nobody around, but the doors were open, so I walked inside, finding myself in a large, empty room, with a grand piano and a teak bar. It was gloomy and I was already feeling a little unnerved when I noticed a pale-faced man standing in the hall. He was dressed in fatigues, with a rifle slung over his shoulder. Behind him, in the half-darkness, there were other men and women, standing against the walls and reclining on rattan chairs. They wore formal dress, as if they were about to be called to dinner. British officers, colonial officials and their wives, their hair was neatly combed and their expressions serene. Beyond them, I came upon more British soldiers, this time in uniform. Crouched behind imitation rocks in a surreal recreation of a jungle landscape, they were aiming their rifles along a teak-panelled wall. The waxworks might have been creepy, but they had been carefully made, and the Governor's House had been meticulously restored. The overwhelming impression was, improbably, of nostalgia for colonial rule.

This didn't make any sense. Why would the generals, who had spent decades railing against colonialists, accept a reimagining of colonial history that not only commemorated but actively promoted the British occupation as a kind of fairytale?

There was nobody around to ask, so I took a seat in the dining room and looked up the hotel's website. 'Enjoy the elegance and splendor of bygone days and traditional luxuries in a unique colonial style,' the website said. It described the house as 'stately… the perfect location where the Governor of Burma chose to rest from the heat'.

Reading on, it became clear that it wasn't just the Governor's House. Other British summer residences had been restored and many of them were now hotels. In the town centre, a clock tower still echoed the chimes of Big Ben and horse-drawn carriages plied the streets. There was a 400-acre park and botanical garden, modelled on London's Kew Gardens, with an impressive collection of orchids, a butterfly museum and an aviary. The regime had even built a large replica of a colonial building to house its administrative offices, painted bright red to match the original brick.

Clearly the British empire had become an opportunity to make money. But wasn't it more than that? When a century of colonial power was reduced to cultural heritage – a collection of beautiful buildings that were marketed as charming – the effect by extension was to soften the image of the elite who now occupied these places. It was a powerful fiction. Here, like in other tourist enclaves in Myanmar, it was difficult to imagine that Myanmar's armed forces were conducting atrocities in the surrounding mountains, in villages and towns that were off-limits to foreigners.

But if this was true, what role did we play by coming to Pyin Oo Lwin, by staying in these luxury hotels, by sanctioning the narrative that everything was fine?

The troubling truth was that we participated in this fiction because it suited us to do so. We still profited from the empire, with our wealth and our freedom. It was exciting to travel here by train and to marvel at how far British power had once spread around the world. It was fun to stay in renovated colonial buildings and to enjoy what the Governor's House website called 'the elegance and splendor' of the past. And it was comforting to imagine that the empire was characterised by its glamour, rather than its brutality, and to admire its material achievements without thinking too much about what had been sacrificed, or what we might owe.

*

Early the next morning, I walked to Pyin Oo Lwin station to catch the train further east into the Shan mountains. Among the crowds on the platform were four soldiers and a tour group of elderly Europeans. On the train, I helped several French women find their seats. They had been staring in alarm at the packed wooden benches in ordinary class and saying to one another: *'C'est pas possible!'* My own seat I gave to an octogenarian German woman, who was taking photographs of everything she saw, while I stood in an open doorway.

Beyond the town, fields of flowers went on for miles. Sunflowers and poppies spilled over the low hills, growing so high that if you walked among them you would disappear. Beside the track, flowers had crushed the fences built to contain them; they even engulfed the train at times. My companions leaned from the windows to pluck them, scattering their seeds.

The landscape became more mountainous as we neared the Gokteik Gorge, an immense chasm in the rock that opened out into a valley more than a thousand feet below. It was bridged by a steel viaduct that was perhaps Britain's most famous contribution to Myanmar's landscape, and was now known among vloggers as the 'most scariest' railway bridge in the world; its imminent collapse had been predicted for decades.

As we approached the viaduct, a Pa-O man – the Pa-O are among the many ethnic groups who live in these hills – offered me some traditional medicine. The perfumed powder was sharp and bitter and made my tongue ache and my throat burn, but it also gave me the confidence I needed to travel across the rickety viaduct.

Suddenly, everyone was pulling out their phones and selfie sticks and rushing to one side of the carriage to get the best view, and we were all leaning from the windows and looking down, terrified, into the forested ravine below. Far beneath us and half-concealed by jungle there was another railway bridge that shadowed our passage across the gorge. Gokteik was a vital link between northern Shan and the central plains, which made it highly vulnerable to attack. The entire area was defended by maximum-security bases and the surrounding jungle was laced with landmines. But the generals still feared the viaduct would be blown up and had built a second bridge for emergencies. Like the viaduct, it was in a terrible condition: rickety, overgrown and clearly unsafe.

If Gokteik symbolised the empire in Burma, the lower bridge represented its painful legacies of conflict and paranoia. Neither bridge led anywhere. Not long after the viaduct opened, the Indian government ordered all work on the project to stop. It was as a waste of government resources, the Viceroy said, and he described the plan to extend it to China as an episode of 'midsummer madness'. The railway was abandoned at Lashio, a remote town that once marked the north-eastern edge of the empire in Burma.

The British never really consolidated control over the surrounding territories, despite claiming them on their maps – and neither had Myanmar's armed forces. Even this railway that tethered these mountains to the central state would not hold. Within a few years, in 2021, Myanmar's military would stage another coup, overthrowing Aung San Suu Kyi's government, killing thousands of pro-democracy protestors and sparking a renewed civil war. Armed ethnic groups and Burmese militias would advance from the hills towards the central Burmese plains, capturing Gokteik and then closing in on Pyin Oo Lwin.

I was travelling on this dead-end railway in a brief period of optimism and openness. But now as I write this, more than a century after the line was built, the tourists are long gone and Myanmar is closed once again to the outside world.

Beyond the bridge, we entered another realm of flowers and then the train curled around a bend into a tunnel, scraping between walls of fetid soil. We entered another tunnel and then another. When we finally emerged from the darkness, the sun was so warm on our faces and our arms, and the flowers so striking against the clear blue sky, that my companions let out a collective exclamation of joy.

A note on pseudonyms

Since the 2021 coup in Myanmar anyone who speaks out is at risk. I have used pseudonyms in this chapter to protect people who I met and interviewed during a time of greater freedoms.

The TAZARA Railway, built by China in the 1970s, is an iconic and politically contested route connecting central Africa's copper mines with the Indian Ocean.

THE FREEDOM RAILWAY
SAM WILLIAMS

From the Swahili Coast towards the Copper Belt

I arrive at Dar Es Salaam station at 10 a.m. accompanied by the explorer Dwayne Fields and a small documentary film crew. We are to travel four thousand kilometres across Africa, from coast to coast, starting on the TAZARA Railway line. The service towards Zambia is scheduled to depart at eleven but the place is in a stupor. The vast concrete forecourt is deserted. In the atrium of the main terminal building, families rest among an archipelago of luggage. There are boxes and bags, caged animals, mounds of fabric, bicycles and countless small children. Heat drifts sluggishly in the air.

Alarmingly, there is no sign of a train.

Locals appear to be using the tracks as a shortcut. A man picks his way across the first track, hoists himself up onto the central platform, walks across it, jumps down onto the second track, and vanishes again into tall grasses, beyond which is the corrugated metal roofscape of a shanty town. I decide to go for a wander too. As I slip off the platform, I nick a small chip out of my wedding ring. Angry with myself for my clumsiness, I climb straight back up and return to the cars that had brought us here from the hotel.

Willy, our fixer, makes some enquiries. A sullen teenager masquerading as a station official tells us that the train is delayed. But he doesn't say by how long. Someone nearby claims that it will only be an hour; another that it will depart at 3 p.m. the day after tomorrow. We don't know who to believe. Willy is as clueless as we are. A whiteboard placed outside the front of the station seems possibly to serve as an announcement screen, but it is as illegible as a tablet of hieroglyphics. And anyway, Willy explains, 'Swahili time' bears only a coincidental resemblance to normal time. It is illuminating that many of the words meaning 'hurry' in Swahili have Arabic, rather than Bantu, roots.

Sunlight bends gently over the course of the afternoon. Shadows of the honey-sweet frangipani trees outside steadily lengthen and sharpen. Planes take off and land at the airport just over the fence. We doze in the cars, watched by bored coffee-sellers. A bus full of Chinese tourists hurtles briefly into the precinct, stops for photos against the station façade and zooms back out again. The more you look, the more you notice the oriental influence of the building. If it wasn't for the tropical birdsong and hot African sun, Dar Es Salaam railway station wouldn't look out of place in Tartary or Tibet. It is like a Maoist mausoleum.

Eight hours after we arrived, there is a sudden commotion. As if responding to a subliminal signal, everyone who has been hibernating around the station springs into life. Panicking, we grab our bags.

The platform is pandemonium. For the past five or six hours, hundreds of people have accumulated in the waiting area from which several flustered pigeons have spent the afternoon fighting vainly to escape. In some parts of the world, a big crowd waiting unaccountably for half a day would have flared into something ugly. Tanzanians, however, have exemplary patience and tensions have been expressed in hangdog faces rather than boiling tempers. But now all that has changed. The levee – which took the form of a puny guard in an ill-fitting uniform – has ruptured. Anxious passengers are flooding onto the platform, the sea of faces fixed with concentration. Two trains have arrived simultaneously. No one seems to know which is which. Willy is frantically trying to apprehend someone to ask, but no one is prepared to jeopardise their seat or berth by stopping to answer our questions. Gradually, it becomes clear that most passengers are boarding the train on the left-hand track.

Willy is going methodically around the group, giving everyone a hug and handing us over to his apprentice, Joshua, and to Freddy, a grizzled immigration specialist who will oversee our passage into Zambia at the other end.

The platform is draining and the train is filling up. The engine grunts splenetically. The horn issues a deafening blast. Our bags and boxes are now piled on the concrete. Someone suggests that we pose for a group photo, but there is no time. We form a human chain, loading everything into a carriage that we don't even know is one of ours. With seconds to spare, we are all on.

The train starts moving, imperceptibly at first. It creaks and groans, shudders and shakes, shrieks and squeals. It is like a statue, miraculously animated, struggling to control its first jerky movements. There is a palpable sense of excitement – a tense energy. Willy, still standing on the platform, his hand raised in salute, is getting smaller and smaller. Now, the platform is behind us, the two tracks have become one and an immense journey has commenced. But you wouldn't necessarily know it: there are no maps, no announcements, no sign of any staff. We shunt our things into the cabins that Joshua tells us are ours. Tinny gospel music plays next door to mine and Dwayne's. It has four bunks, each of which is equipped with a plastic mattress, greasy-looking pillow and a rolled blanket. Kids career up and down the corridors, overcome with excitement. Before we are out of the station, opportunistic traders are already knocking on doors trying to sell beers and bananas.

Had we left on time, we would have spent the first afternoon admiring the glorious plains of the Nyerere National Park – an arcadian landscape, rich with wildlife and acacia trees.

As it is, the light is fading fast.

Dusk settles heavily on the suburbs of Dar Es Salaam. The world is evanescent. A forest of heads and arms crowds the open windows along the length of the train. The immediacy of the environment is astonishing. The tracks slice directly through a community coming alive at the end of a hot day, mere inches from people, houses and trees. I could knock hats off people's heads as we pass. We see a game of football, a church service, herds of motorbikes, a *muezzin* peering at a Qur'an in a minaret, women carrying baskets of nuts and tubers on their heads, busy night markets. Instantly, the salt-sprayed and spice-seasoned Swahili Coast has been replaced by something more obviously African – a burnt-orange sky, the smell of woodsmoke, scrambled music, vultures silhouetted on treetops.

Night has almost fallen. The air outside, now rushing past the gaping windows, is chilly. Passengers retreat inside.

Windows slam shut. People are getting ready for a forty-eight-hour odyssey. The train turns slightly, arcing south. The last of the sun catches the track, like an ember spat from a fire. The metal seems for a moment to light up, marking a path of steel into the heart of Africa. Our journey to Angola has begun.

*

The myth of the TAZARA Railway began almost a century before it was completed.

Cecil Rhodes envisaged it as a section of an ambitious rail route connecting Cape Town, on the windswept southern tip of Africa, to Cairo, almost within sight of the Mediterranean. The Cape to Cairo railway would be the armature upon which Great Britain's African empire would rest. It would bring English manners and morals into the heart of the 'Dark Continent', fortify her colonies against encroachment by European rivals and unlock enough raw materials to keep Wales's foundries and Lancashire's mills employed for another hundred years. It was a big dream, but not an unrealistic one. Much of that line *was* built while the territory it crosses was under colonial control – but the segment connecting Zambia with Dar Es Salaam was not.

Zambia gained independence from Britain in 1964. Until then, it had been called Northern Rhodesia. The open pit mines of its Copperbelt Province were the country's economic dynamo. Since copper production had begun in the late nineteenth century, the principal export route was south. A network known as the Rhodesia Railway, which was jointly owned and run by the colonial authorities in

Northern Rhodesia, Southern Rhodesia (later Rhodesia, then Zimbabwe) and Bechuanaland (later Botswana) was built to carry minerals from Copperbelt to Indian Ocean ports such as Durban, in the British colony of Natal (now part of South Africa), and Beira, in Portuguese Mozambique.

Zambia is landlocked. Its economic fortunes depend to an uncomfortable extent on the behaviour and policies of its neighbours. As the new Zambian flag rose above Lusaka at midnight on 23 October 1964 (applauded by, among others, a heavily bejewelled Princess Mary, the Princess Royal), it signalled the beginning of majority rule in Zambia and the latest step towards a free Africa. But five-hundred kilometres to the south, in Salisbury, capital of the British colony of Southern Rhodesia (which, after Zambia became independent, began to refer to itself simply as Rhodesia), a rival version of history was being asserted. In a continent gusting through with the 'winds of change', Rhodesia was grimly determined to resist. Aggrieved at the speed with which other British territories were being liberated and yet unwilling to compromise its cherished principle of white-minority rule, the Rhodesian cabinet had begun preparing a dramatic measure. In November 1965, only a few days after the first anniversary of Zambia's celebrations, it made a unilateral declaration of independence.

It was a gambit that precipitated a crisis in Zambia.

Zambia's first president was Kenneth Kaunda, a charismatic nationalist who made no secret of his distaste for the renegade Rhodesian government. Yet he was also conscious that Zambia was, due to its inherited infrastructure, reliant on its cooperation. Virtually all Zambia's precious copper exports had to pass through its southern neighbour to reach the sea, crossing into it via a heart-stopping steel bridge just yards downstream from the thundering Victoria Falls. Kaunda didn't want Zambia's economy to be hostage to an unpredictable and antagonistic regime. He thus identified the creation of an alternative route to market, ideally through friendly states with black-majority rule, as a strategic priority.

The Benguela Railway, which ran to the Atlantic Ocean from the copper-rich Congolese province of Katanga, was examined as an option. The Benguela Railway had operated between the Atlantic port of Lobito and Katanga's mines since the late 1920s. In fair conditions, its eucalyptus-powered steam engines could make the journey in less than two weeks: much faster than the journey from Kapiri Mposhi (the main railhead of the Zambian copper belt) to Durban. But the Benguela Railway was an unsuitable insurance policy. First, Katanga itself was riven by fighting in the aftermath of its own attempted secession from Congo in 1960; and second, further west, Portuguese paratroopers were at that very moment waging a brutal bush war to retain Angola as a European settler colony. Substituting an export route through Rhodesia for a route through Congo and Angola offered Zambia uncertain benefits.

As early as 1961, Kaunda and his Tanzanian counterpart Julius Nyerere had discussed the possibility of reviving the long-lost idea of a railway linking Copperbelt directly with Dar Es Salaam. This time, they determined that it would be built on their own terms, to serve their own development, rather than as a tool of outside exploitation. In a speech several years later, Kaunda explained that it was imagined in those early conversations as being 'a freedom railway, a railway for

strengthening African unity and independence … [and not]
an instrument of imperialism and colonialism'.[1]

While Zambia planned its independence celebrations in
the autumn of 1964, Tanzania was busily forging its own path
as a sovereign state. It had secured independence three years
earlier, in 1961, and took its present form in April 1964, when
Tanganyika and Zanzibar merged. Like Kaunda, President
Nyerere was a trenchant nationalist. He was a visionary and
idealist who had lofty aspirations for Tanzania to become a
new kind of African polity: progressive and egalitarian, with
a socialist tilt.

In 1961, Tanzania's relationship with China was a history
of tangential encounters.

The Swahili Coast had been visited by Chinese travellers
in earlier centuries. Emissaries from Beijing were welcomed
periodically in the *majlises* of Zanzibar's sultans. In 1431, the
King of Malindi, an island just north of Mombasa, sailed
with his court to Bengal. There he presented Admiral Zheng
He – the famous Ming Dynasty explorer – with a giraffe. The
creature was taken back to the Forbidden City as a gift for the
Yongle Emperor, who is reported to have let it roam among the
orchids and lotus ponds, explaining to astonished visitors that
it was a unicorn. In 1910, the German colonial government
in Dar Es Salaam enlisted the services of Chinese workers to
build its own railways, which served plantations in the north
of the country. But in general, as European penetration of east
Africa deepened, and China itself sunk into a mire of civil
war and social breakdown, contact remained sporadic and
superficial.

---

1   Martin Bailey, *Freedom Railway: China and the Tanzania–Zambia Link*, Collings
(1976), p. 15

This changed in 1964.

Zhou Enlai, China's chief diplomat and the second highest-ranking member of the Chinese Communist Party after Chairman Mao, embarked on an African 'safari'. The ten-country tour was a chance for Zhou to drum up African support for Beijing's claims to legitimate statehood (which were being noisily contested by the rival nationalist government in Taiwan). It was also an opportunity for him to earmark potential Chinese investments in Africa. The safari culminated with a speech in Accra, Ghana. During a banquet at Osu Castle, a whitewashed fort with battlements dashed by Atlantic surf, Zhou stood to thank his host, President Kwame Nkrumah. He also articulated 'eight principles of Chinese aid'. The first of these was that 'the Chinese government always [observes] the principle of equality and mutual benefit in providing aid to other countries'. Despite requiring stilted translation into English, the speech drew eager applause from a roomful of African leaders whose life missions had been to unshoulder the burden of European colonisation, which they believed to be involuntary and of asymmetrical advantage.

Nyerere was at the table. Earlier in the week, he had secured a private meeting with Zhou. The two sparked a warm personal relationship. Nyerere was instinctively open to Chinese overtures. China also seemed to offer a template for how a once-benighted country could, with sufficient will, effort and sacrifice, achieve wholesale social and economic revolution. Despite its enormous size difference (in the 1960s, China's population was bigger than that of the whole of Africa), China was seen by aspirational third world nations such as Tanzania as essentially 'one of us'; rather than, as now, a distant and inscrutable superpower.

From 1962, Tanzania had been receiving limited military aid from China. Detecting Nyerere's receptiveness and spying an opportunity to broaden links, Zhou invited the Tanzanian premier to Beijing for a bilateral summit. Nyerere accepted the offer. In February 1965, among the high-flown surroundings of Beijing's Great Hall of the People, Nyerere and his Chinese counterpart Liu Shaoqi signed a 'Treaty of Friendship'.[2] After a brief but intense courtship, Tanzania had become China's best friend in Africa.

Meanwhile, discussions were continuing apace on the railway.

The first practical effort towards building it occurred in 1963 when a British company was commissioned to conduct a feasibility study. The firm, Lonrho, offered to sponsor construction and operate the line. In return, it demanded the right to transport every gram of Northern Rhodesia's copper output. This was unacceptable to Kaunda's government-in-waiting: partly because it would simply have replaced one monopoly with another, doing nothing to diversify the country's economic fortunes, and partly because, due to Northern Rhodesia's joint ownership of the Rhodesian Railway, shifting exports wholesale from the southern route to the eastern route would have required heavy compensation to be paid to Southern Rhodesia.

The World Bank was also asked to conduct a study. The report came out strongly against the railway project. Instead, the authors argued for an upgrade to the Great North Road. The Great North Road was the automotive counterpart to the Cape to Cairo railway, a highway (also first sketched out

---

2   https://www.mfa.gov.cn/eng/wjb/zzjg_663340/tyfls_665260/tyfl
_665264/2631_665276/202406/t20240606_11405573.html

by Cecil Rhodes) that would thread the length of Africa's eastern seaboard. By the early 1960s, the road existed in some form or other from Cape Town to Addis Ababa, but virtually none of it was tarmacked. In monsoon storms it became an impassable bog, earning the moniker 'hell run' among jaded local truckers. President Kaunda was sceptical that an upgraded road would solve his dilemma. The economics of copper demanded something different. He maintained that a new railway would be faster, more secure and – in the long run – cheaper. But the World Bank dug in. With funding and contractors from the United States, it embarked on its upgrade to Zambia's portion of the Great North Road.

The decision to rehabilitate the road did nothing to dissuade Kaunda and Nyerere of the railway's necessity. In fact, it was becoming more urgent. Following Rhodesia's unilateral declaration of independence, its relations with Zambia strained to breaking point. The legacy arrangement under which Rhodesia Railways – the de facto vehicle for Zambia's copper exports – had been jointly owned was moribund, and Ian Smith's troublesome government was looking for ways to wrest full control. Charges on Zambian freight were arbitrarily increased and several brief but disruptive blockades were imposed – for instance, on the import into Zambia of coal that was essential for the running of mines.

In 1965, the Tanzanian and Zambian governments appealed directly to the United States for funding. They were rebuffed. The response from Washington was no surprise: the view previously articulated by the World Bank was a pro forma expression of American sentiment towards the rail project. Uncle Sam preferred the road. In any case, the United States was wary of Tanzania, which it thought was flirting too enthusiastically with communism. Western journalists liked to interpret Nyerere's taste for 'Mao-style' suits as evidence of his politics. A thesis was advanced that the railway was mainly an ideological project, designed to excise the West from African affairs and reorientate Zambia and Tanzania eastward. A report published by USAID, for example, concluded, '...the railway can find no justification except in "unreal" definitions ... the economics of the rail link are clearly not persuasive. What remains are political issues which are often clothed in economic dress.'[3]

American obstinacy frustrated Nyere and Kaunda. They suspected that US arms and cash were helping to prop up the hostile white-minority regimes in South Africa, Rhodesia, Mozambique and Angola. They also observed America's lavish spending elsewhere. One Tanzanian official lamented that '...our railroad could be built for what the Americans are spending in Vietnam every four days'.[4]

A request submitted to Britain received similarly short shrift. At the time, Britain was suffering an acute balance of payments crisis and frantically cutting its international aid budget. It was also reluctant to lock up capital in long-term projects in countries that had only recently kicked it out. In 1963, during a visit to London, Kaunda discussed the project with Secretary of State Rab Butler, but no financial commitment was forthcoming.

And so China stepped in.

In June 1965, only four months after the Treaty of Friendship was signed in Beijing, Zhou Enlai travelled to Tanzania. Archive footage shows him being welcomed by

3   Bailey, p. 44
4   Bailey, p. 46

President Nyerere on the asphalt at Dar Es Salaam airport. The scorching sun of the Swahili Coast hammers down on an honour guard of troops and tribal dancers. Nyerere presents Zhou with a garland of flowers. The greying revolutionary wears it around his neck as he and Nyerere stroll to the terminal, looking unusually dandyish as he greets well-wishers and reporters.

In return for the floral tribute, Zhou told Nyerere that China would be willing to build and finance the coveted railway. He described it as the 'friendship route'. It was a seismic moment. After Tanzania and Zambia had been trying and failing for years to find a Western sponsor and after having been told repeatedly that the railway was an illogical folly, the intervention from China was as invigorating as a lightning bolt.

China was still, by any measure, a poor country. At the core of its proposal was a loan of almost two hundred million dollars. It was by the far the biggest foreign aid package China had ever advanced; a huge statement of political chutzpah. After the Aswan Dam in Egypt and the Volta Dam in Ghana, it was to be the biggest infrastructure project ever undertaken in Africa. Despite tropes about Nyerere being a communist, he and Kaunda were wary of China's motives. Their dogged attempts to win Western backing suggests that their preference – albeit a conflicted preference – was for American or British support. But the Chinese offer was too attractive to ignore. Moreover, it seemed to come with none of the onerous and pedantic conditions that the others had demanded. China would provide as much manpower, money and material as needed.

China's offer was not immediately accepted. Both African parties wanted more time to consider their options. News of

China's move caused a spasm in Western media and among Western governments. American and British diplomats went on the offensive in Dar Es Salaam and Lusaka, warning their hosts of the risks of doing business with China, and trashing China's technical expertise and capabilities.

Yet, despite the flurry of activity that Zhou's offer provoked, no acceptable counter was produced. Nyerere complained that '…the West, which doesn't want to build our railway, doesn't want China to build it either'.[5]

Thus, in 1967, a delegation of Tanzanian and Zambian officials travelled to Beijing to sign an initial agreement for China to finance, build and equip the railway. Three years later, in 1970, the final agreement was ratified. Joined by Chinese and Tanzanian dignitaries, Zambia's President Kaunda laid a foundation stone in the Dar Es Salaam suburb of Yomba, which was eventually to germinate into the terminal in which we had just spent an afternoon. Within a few days, Julius Nyerere laid a foundation stone in Kapiri Mposhi, over a thousand miles away at the other end of the line. Both ceremonies were snubbed by Western diplomats.

That was no matter, as far as the African nationalists were concerned. The West had had its chance and wasted it. Finally, the long-held dream of connecting Copperbelt to the Swahili Coast, freeing independent Africa from the influence of the white-ruled south and breaking once and for all the economic stranglehold of history, was about to come true. The great *Uhuru* (or Freedom) Railway would be built. And it was to be done with China's help.

---

5   Bailey, p. 45

*

On the train, things settle quickly into a rhythm.

Everything sways and bobs. The train is like a caterpillar, with each carriage and vestibule moving in a different direction, as if it is trying to shake something off. Walking down the corridors demands the balance of a gymnast. With every unpredictable motion of the train, one risks toppling through the open door of a crowded cabin. I think back to the narrowness of the tracks at Dar Es Salaam station. How can such a big, beastly machine stay upright? The gears and wheels clip out a hypnotic beat, like a well-drilled cavalry platoon on interminable parade.

We assemble in the dining car. It is dimly lit and clinical, with the atmosphere of a roadside diner crossed with a doctor's waiting room. Its plastic seats are easy to wipe clean and the windows – which don't open – are glazed from the inside by decades' worth of beady condensation. I try to look out, but all I see is my featureless reflection.

In an adjacent kitchen car, a team of women bustle around a steaming row of pots sat precariously on a bank of stoves. I watch broth slopping over the brims with each tilt of the train. The cooks aren't fazed. A waitress comes to our table with a tray of Serengeti beers, which we receive gratefully. The food options are limited: meat and fufu, chicken and fufu, fish and fufu, and plain fufu. We take our pick.

As we wait for our meal, we listen to Freddy's stories. His father was a chief in the Kenyan borderlands and the proud holder of an OBE. Freddy is tall, well-fed and has an aristocratic bearing. He has salt-and-pepper hair and a beard as wispy as a Sufi saint's. He travelled on the TAZARA as a younger man and remembers the feverish anticipation when it opened. It was an augury for the future of a newly independent Tanzania. The country had never felt so modern, so free, so connected.

Tanzania is a country prone to idealism. In 1967, six years after independence, Julius Nyerere gave a famous speech in the northern city of Arusha which outlined the precepts of a specifically African form of socialism – or *Ujamaa*, which means something like brotherhood in Swahili.

What became known as the Arusha Declaration encoded in Tanzania's politics a perfect form of classless egalitarianism. With stormy rhetoric and wild gesticulations, Nyerere willed a deeply tribal society that had been pummelled and distorted by two hundred years of colonial rule to somehow transcend its manifold challenges and become a case study in a political theory textbook. While Zambia's motivations for the railway were always, at root, pragmatic, the TAZARA project fitted into Nyerere's more mystical theory of Tanzania's destiny. Not only was it to be built by China, a Marxist friend exorcising its own colonial demons, but it promised to open the country up for the first time, dissolving barriers between those in the cities and those in the rural hinterland. Like railways everywhere, it had a providential narrative. It would be a vehicle of change and an instrument of progress. Intervening years may have discredited the Arusha Declaration as a template for government (if not as a philosophical ideal), but its spirit still survives in the swaying onward movement of this locomotive, fifty years later, which is transporting a thousand or so travellers into the African night.

As Freddy speaks, he is interrupted by a succession of friendly passengers stopping at our table for a chat. Each is more intoxicated than the last. There is Thomas, who is on his way to an aunt's funeral in Makambako, accompanied by a gaggle of sisters who we can see peering bashfully around the door to the dining car. He insists on buying us more beers, but – for his sake and ours – we signal 'no' to the waitress when he turns his back. Thomas is followed by Ibrahim, a retired TAZARA driver who, as far as we can tell, now spends his days haunting his old workplace – unable or unwilling to countenance a sedentary life. And there is a man whose name we don't catch, so drunk he almost collapses onto our table when the train slews sharply over a warp in the track. In slurred Swahili, he asks where the toilet is.

Freddy looks around.

'Go up there,' he says, waving vaguely towards the next carriage.

The drunkard staggers into the vestibule. In full sight of everyone in the dining car, he clumsily unzips his trousers and starts urinating on the floor. He groans with relief. His eyes are closed and his knees are about to buckle. Dwayne can barely contain his laughter. An outraged waiter pushes past us, angrily berating the man for violating the decency of his restaurant. Wisely, Freddy has vanished. We later see the villain being carried gently into a cabin by two of the on-board police officers, his trousers still damp and his body hanging limply in their arms like a puppet following a performance.

Later, we try to sleep. It is a strange and disorientating experience as the train lurches its way across the darkened plains. Occasional stops yield unsettling silences. There is no light except starlight. At one resting point, the train is perfectly still and quiet. As I lie on my bunk, my ears pick out the footsteps and breathing of someone, only a metre away, creeping along the tracks in the blackness.

We wake to the sounds of an impromptu bazaar blooming alongside the stationary train. People are jostling up and down, shouting into the windows, trying to hawk water, soda, bananas, sugar cane or fried dumplings. According to the sun-bleached sign on the crumbling platform, we are in Mchombe. Mchombe seems to exist only to serve the trains that pass infrequently through, like an African Adlestrop. I look up at the 'high cloudlets in the sky'.

The solid ground beneath my feet is unnerving. The soundtrack is that of a typical village morning: disembodied shouts, cocks crowing, footsteps in the brush and the clatter of decrepit bicycles through banana groves. Monkeys patrol the treetops. The light is viscous and mellow, like burnished gold. In either direction the train track snakes into thick bush.

Word has spread that we will be staying in Mchombe for an hour or two. I decide to scout along the tracks. I step over them and step back. It is an oddly powerful feeling to fit what is supposed to be one of Africa's mightiest pieces of infrastructure between my feet.

The driver has chosen a spectacular place to stop. Jagged green mountains cascade along the horizon. Ahead, the track forks. One branch continues towards the mountains and the other seems to come to a halt in a shed out of sight of the train. Using the sleepers as stepping stones, I follow the tracks to the train shed. The shade provided by its corrugated iron roof is merciful. Thick weeds writhe up from cracks in the concrete and rusted tools lie scattered like neolithic weaponry in a Gloucestershire field. If the building had been empty, I would have concluded that it had been long abandoned to the elements. But there are two huge, hulking engines in here, painted the same colours as the machine that has tugged us from Dar Es Salaam. They are streaked with rivulets of long-dried rain. But they also look Herculean, as strong as fire-powered pack animals. They are great lumps of metal-muscle with no purpose other than to pull dozens of incredibly heavy carriages across the highlands of East Africa. If the dining car is the pleasure palace of the TAZARA Railway, these engines are pure business.

I spot someone in the pit beneath one of the engines.

'*Jambo*! Hello! Do you speak English?'

The mechanic hoists himself out, dusts off his overalls and says that he does. I ask him what he is doing.

'These trains are very old. They always broken. I make sure they can keep going until Makambako,' he answers.

'Are they Chinese?'

He says yes. And then: 'I think they will soon come back to replace them.'

This strikes me as a good filming opportunity, so I use my two bars of signal to call Léo, the director. The mechanic gets back to work. I sit on a buffer staring out the shed. The main train is obscured by tall grasses that are vibrating with insects. But among the villagers walking along the shimmering tracks, three familiar figures are taking shape. As they get closer, I can see that they are jogging – the clumsy, laboured gait of people carrying expensive camera equipment in the hot sun along a path of treacherous wooden planks.

They arrive and flick on the cameras immediately. César, the cameraman, walks around one of the engines, scanning it, capturing it from all angles. Meanwhile, Léo is recording Dwayne, the presenter, in conversation with the mechanic.

They are in the cabin and the mechanic is explaining what each of the knobs, dials and levers does. The dashboard is stencilled with warnings and commands in Chinese, Swahili and English. The cab window is crusted with dirt and splattered bugs.

As I am standing back, trying to stay out of shot, I notice another figure moving towards us. It is Joshua. He is panting and sweating. Before he gets into the shed, he shouts that our train is about to set off.

Dwayne, Léo and César are in the cabin and cannot hear. I run around and gesture vigorously. 'THE TRAIN IS LEAVING,' I mouth.

The message gets through. In panic, they say a hurried *asante sana* to our impromptu expert and descend the narrow metal ladder, jumping the last metre-and-a-half onto the cracked concrete platform. The cameras are lowered and we start to run. Joshua has stolen a lead but we soon catch up. The train's horn is shattering the tranquil village morning, the driver making repeated pulls on the metal chain dangling from the ceiling of his cabin. He seems to be enjoying it.

The locomotive comes into view. People are returning to it along its full length, pulling others up behind them and calling out to folk left behind. Alexandre, the expedition photographer, is near the front of the train, on the platform, hurrying us up. We are parallel with the engine now but still have a long way to sprint before we can get on. The platform is an obstacle course of market stalls, piles of timber, sacks of grain and flour, buckets of pomegranates, parked motorbikes and people standing chatting. We dodge them, their colours blurring, the horn still shredding the air. Faces crowd the windows of the third-class cabins, laughing and pointing at the hilarious sight of a group of *muzungus* racing for their lives.

The train releases a huge hiss and, with a primal groan, it starts to move. Our carriage is at the back of the train, but we are still only halfway along. We can't risk it. One of us jumps on the ladder leading into a random vestibule and heaves himself up. The cameras follow, with the others coming up behind. We are aboard. Drenched in sweat and laughing manically, we are still on our way.

Every stop is similar. An entire town descends on the station, everyone desperate to sell something to the hungry, thirsty and bored population on board. Some towns are big, with buses, cell towers and concrete buildings. Others are mere hamlets; little more than name signs planted among the huts composed of plywood and tin and stuck together with dung and mud that line the track.

In these places, it is unclear what value the railway has provided. When it was built, the export of Zambia's copper was its key commercial justification; but it wasn't the entire story. In Tanzania, the track was supposed to bond the rural periphery with its urban centres. It was supposed to unearth fertile agricultural lands in the south-east of the country that had been neglected by German and British colonisers, who preferred to grow their sisal, cotton and tobacco in the centre and north. It would accelerate the country's journey to self-reliance. Even if it was never likely to trigger the profound metamorphosis that China had experienced, it would at least help to reduce the country's widespread poverty.

Fifty years later, it doesn't seem to have done so. Whatever seeds of commerce or industry were sown along the track have failed to grow, unless you count opportunistic sales of cola and splintery tubes of sugar cane. There are no factories, no offices. Peasants continue to till the fields, digging up just enough to eat. Marxist ideals are nowhere to be seen; at least not through the windows of my cabin. Nyerere's vision of an affluent, classless society remains a chimera.

Indeed, the villages we pass through are much as they would have been before the Chinese worker gangs churned their way through in the early 1970s; before Tanzania won its independence in 1961; before Great Britain expropriated 'Tanganyika' from a defeated Germany at Versailles in 1919, like a neighbour claiming a bargain at a yard sale. The essential structure of life here has changed little since David Livingstone had his lakeside meeting with Henry Morton Stanley in 1871, or since caravans of Arab merchants passed through in the fifteenth century, carrying gold panned in the glittering shallows of the Zambezi River. The train track – which, when viewed at a macro scale, is part of a cutting-edge story of global trade and power – seems to have much the same status in these places as a stream or forest trail. It is something that is there, that to all intents and purposes has always been there, but which is not worth thinking about much, useful only as a shortcut up the valley.

*

The landscape evolves. We are gaining altitude. The lush green plains give way to more spartan uplands. The train crosses several dizzying gorges filled with raging water. A series of tunnels, one after the other, throw the train into absolute darkness before flooding it again with light. The effect is as disorientating as a solar eclipse.

The construction of the TAZARA involved an estimated fifty thousand Chinese workers. They were joined by a hundred thousand locals, representing almost two per cent of the male population of Tanzania and Zambia at the time.

The countryside was trawled for recruits. Unemployed young men flocked to it. Hiring posters made it clear that candidates must be literate. They also needed to have completed primary school, a criterion that weeded out most applicants. The goal of both the African and Chinese governments was for the railway to eventually be self-sufficient, although China did promise to keep a cadre of technical experts on the ground after the handover. The authorities therefore sought workers in vigorous physical and intellectual health who could, with a bit of coaching, become bona fide managers and engineers. Formal training centres were opened in Mang'ula, Mgulani and Mbeya in Tanzania and at Mpika in Zambia. At least two hundred Tanzanian and Zambian students were even sent to Beijing's Jiaotong University: a specialist institute whose stern emblem features an anvil, a hammer and a chain. There, they learned to dismantle and reassemble trains, and practised their skills as conductors on real Chinese railways with real Chinese passengers.[6]

Their Chinese colleagues were, by and large, skilled and experienced.

Many were former People's Liberation Army (PLA) engineers. Like the US Army Corps of Engineers, which

---

6   Liu Haifang & Jamie Monson, 'Railway Time: Technology Transfer And The Role Of Chinese Experts In The History Of TAZARA' in Ton Dietz et al., *African Engagements*, Brill (2011), p. 233

during the Great Depression defibrillated America's dying economy by building dams, ports and airstrips, the PLA was used by the Communist regime in China to subdue, control and reshape its vast and complex landscape. Many of the workers arriving in Dar Es Salaam in 1970 had only recently been relieved of duties on the Chengdu–Kunming Line, a seven-hundred-mile trunkline through the fantastical limestone mountains of Yunnan and Sichuan. China had a widely acknowledged – if not always respected – heritage of railway engineering. Not only were Chinese labourers employed on the colonial railways in the Belgian Congo and German East Africa, but thousands contributed to the construction of the railroads that connected America's Atlantic and Pacific coasts in the nineteenth century. The men China deployed to Tanzania and Zambia between 1970 and 1975 were far from the buck-toothed coolies pilloried by the country's opponents. They were up to task.

Equipment could not be transported on the rival lines operated by the Rhodesia Railway or Benguela Railway. Instead, it had to be carried on trains grinding slowly along completed sections of track. Some smaller things were driven in trucks on those sections of the Great North Road that ran parallel to the TAZARA route. The road was being refurbished at the same time as the TAZARA was being built, financed by the United States and with lots of American contractors. The irony wouldn't have been lost on those involved that the American road was being used to help build the Chinese railway.

During the construction period, Tanzania's economy was warped by the TAZARA's awesome gravitational force. The sea off Dar Es Salaam was dotted with ships waiting to dock and unload trucks, machinery, steel, cement, locomotives and rolling stock. Dar Es Salaam's port capacity was swiftly gobbled up and new quays had to be built. Meanwhile, in 1973, complaining that Zambia was harbouring guerilla fighters who were using the existing south-facing rail link as a sort of Ho Chi Minh Trail, the government of Rhodesia closed the bridge across the Victoria Falls. This was a blow to the Zambian economy, which could now only export its copper via trucks. A few years earlier, such a move by the Rhodesians would have forced almost any concession from Zambia, but with the TAZARA proceeding towards Copperbelt Province at breakneck speed, President Kaunda held his nerve.

The line is split into distinct topographical sections. First, on departure from Dar Es Salaam, it chugs southeast across enormous grasslands. Next, it confronts a dramatic and thickly forested rockface. This is the Mufindi Escarpment – part of the Great Rift Valley – and the access point for the plateau beyond, which stretches almost all the way to the Atlantic on the other side of Africa. At the top of the ridge, near the town of Mbeya, the TAZARA reaches its peak altitude of just under six thousand metres. From there, the journey to the terminus at Kapiri Mposhi is long, flat and straight.

The stretch of track leading up to Mbeya – where the line climbs the fearsome Mufindi Escarpment – covers only eight per cent of the TAZARA's total length. But it is home to most of its three hundred bridges and twenty-one tunnels. It was this section that gave birth to the TAZARA legend.

As a piece of realpolitik, the TAZARA was audacious. As a piece of engineering, it was a marvel. It is difficult to overstate how hostile the terrain was through which it was built. Today, the tracks have given rise to a smattering of settlements, complete with feeder roads and power lines and occasional bursts of 3G mobile signal. It is difficult to imagine the landscape without the railway. But in the 1970s, the area where the lowlands swelled suddenly into highlands was a raw, predator-spooked wilderness. The only tracks were those of animals and people and the only sounds were the whispering of the wind and the lonely screech of raptors.

The advance parties of Chinese surveyors had to be protected by armed scouts. Most nights spent in tent camps would be interrupted by gunshots and the bone-chilling screams of wounded beasts. It was not the first time a rail project in East Africa had been plagued in this way. In 1898, a pair of lions stalked the construction camps of the Uganda Railway, near Kenya's Tsavo River. Two culprits were eventually shot, but not before they had devoured many African and Indian workers. The marksman was a British Army officer and engineer called John Henry Patterson, who had been a prolific tiger hunter during earlier military service in Punjab. Patterson kept the skins of the infamous creatures as rugs in his home until, in 1924, they were sold to the Field Museum in Chicago. There, they were reconstructed and turned into a diorama. School groups and tourists still gawk at them today: laughing, taking photos and perhaps experiencing just the faintest flutter of fear that the crouching man-eaters might suddenly snarl back into life.

Blasting, scouring and planing a route through this forbidding range demanded grit and heroism. Surprisingly few deaths were recorded, but there were lots of injuries and the flesh of the workers was feasted upon by mosquitoes and tsetse flies. Unlike earlier colonial-era projects, the construction of the TAZARA was relatively egalitarian. The last of Zhou Enlai's eight principles of Chinese aid stated:

> ...the experts dispatched by the government of China to help in construction in the recipient countries will have the same standard of living as experts of the recipient country. The Chinese experts are not allowed to make any special demands or enjoy any special amenities.

The burden of labour was shared. Chinese foremen were in charge, and they were more likely to roll up their sleeves and get mud on their faces than direct proceedings from cushioned camp chairs beneath parasols. Local recruits were amazed by the industry of their Asian colleagues, who maintained a round-the-clock tempo for five years. In one grainy archive clip, a platoon of bare-chested Chinese and African workers are up to their navels in torrential water passing through a cramped tunnel. Their faces are streaked with sweat and furrowed with fatigue. With no regard to hierarchy, they are bailing out the clay-brown liquid as if their lives depend on it.

The TAZARA Railway (which was initially called the TANZAM Railway, before being renamed TAZARA in 1984) was finally completed on 23 October 1975, when the first train setting off from Dar Es Salaam rolled triumphantly into the sidings at Kapiri Mposhi. Initial estimates had suggested that it would only be finished in 1977. It was two years ahead of schedule and almost within budget.

The sheer number of Chinese workers involved in the TAZARA project, plus the daunting scale of the technical challenge, meant that it was quickly co-opted by the Chinese state media as a parable of national valour. It has maintained this status ever since, evidenced by the Chinese tourists we saw paying homage at the terminal in Dar Es Salaam.

The propagandists weren't without justification.

Since the 1970s, China has financed and built many bigger projects. But few have the same magnetism as the TAZARA. It tells the story of a turning point in history, when China finally became a great power. It seemed to prove China's own redemptive rebirth from crumbling empire to modern superpower. It defied its many doubters and critics, who had traduced Chinese capabilities and predicted that the project would fail. And it fashioned a basic template for Chinese action overseas that still applies today. Despite its powerful immediacy – the shrieking brakes, the smell of dried sweat, the howling axels and the cold air clawing at cracks in the windows – the TAZARA is steeped in nostalgia. It is as much a memory as something that still exists.

*

At twilight, we reach Makambako. The evening light is coppery and cinematic. A vast prairie dotted with *kraals* spills towards every horizon. Vistas like this cannot but inspire a sense of yearning. It is hard to believe that only two days ago we were standing on a beach watching mango-wood dhows bob about in turquoise surf and listening to the *adhan* drifting on the breeze.

Finally, in the middle of the second night, we pull into our destination: Mbeya – a town known, according to Freddy, as 'the Scotland of Africa'.

As the train approaches the station, its horn issues a final deafening clamour, violating the chilly silence. For reasons no one can properly explain, the service is not travelling all the way to the Tanzania–Zambia border, let alone to Kapiri Mposhi. From here we will need to drive the rest of the way to Zambia. Luckily, we had known this before we departed. But some of our fellow passengers weren't told. As news spreads that the train will not be continuing beyond Mbeya, a sense of confusion and resignation washes over the train. We hear wails and shouts. The bus station will be busy in the morning.

Massimo Vignelli in his apartment signs a copy of his 2016 *Vogue* map for Ovenden.

MISADVENTURES IN MAPPING
MARK OVENDEN

1.

How do niche interests develop in the young? Track back and you will often find very early experiences are key. I remember a few, if only as fragments: a conversation, through a locked bathroom door; an agitated request, 'Please hurry up, there's a queue forming here,' and from inside, me, 'I'm trying to locate the route of a dismantled railway.' My ingenious (but self-confessedly unorthodox) aunt had wallpapered the interior of her WC with old Ordnance Survey maps.

Here's another: a long journey from one of London's less salubrious western suburbs on the Underground with a frequently fidgety and annoyingly interrogative child who is instantly becalmed by a small pocket map of the system provided by his long-suffering mother.

And another: a tedious car journey; the two boisterous children in the back are silenced by being handed out-of-date road atlases and are trying to spot landmarks shown on the maps.

Or this: A dilapidated industrial landscape; father and son are negotiating the route of a recently abandoned railway, following its former trajectory on a map. I don't remember the weather or what we wore or what we said but I do remember the pleasure in our shared endeavour.

I wasn't just nurturing a fascination with railway transport – I was already specialising – it was the cartography that gripped and, if those memories serve, often calmed me.

Other kids collected Action Man (plus outfits), Dinky toy cars (mostly dented) or conkers (always better than yours)... From as early as I can remember my nerdy assemblage was a motley parade of dog-eared transportation maps. It inspired zero interest or kudos compared to my classmates' stickers of football players, chipped marbles or trump cards (perceived as far more masculine pursuits), and seemed to sit somewhere beneath a stamp collection or autograph book in the contempt my collection earned from other boys. Certainly, by the time I entered 'big' school (a comprehensive) during the early 1970s, my map collection occupied an entire sock drawer (the socks having been rehoused elsewhere, poor things). The tattered transport throng included little from outside Greater London, but then a school trip to Paris changed all that.

We were thirty teenage *enfants terribles* in a French city full of promise. How the teachers managed not to misplace any of us I will never know. Most of the kids were impatient, racing ahead for a glimpse of the Eiffel Tower, tempted by tourist shops and snacks. I was not one of them – I quickly became the straggler the adults struggled to account for as I dawdled, gawping at the wonders of... the Paris Metro. Not only was I (so accustomed to London's Tube) gripped by the shiny door-releasing handles, rivetted by the rubber-tyred trains, spellbound by the screeching door-closing alarms, intoxicated by the intense aromas… but their *map* was just so mysterious, even mesmerising. I was captivated; nothing could have encouraged me to *vive la différence* more.

The mid-1970s Paris Metro map was so utterly unlike the London Underground diagram that I became somewhat fixated by it… and eventually this obsession expanded to those of other cities. The French offering differed to the British one as follows: London's felt ordered, clean, aesthetically pleasing. It had lines that were only horizontal, vertical or of just one angle (forty-five degrees); even the River Thames (the sole reference to surface features) was stylised in this manner. All station names were set horizontally; the central area had been expanded while suburbs were compressed; interchanges were shown using open black circles with white centres – joined to others at larger stations using white-line connectors. The concept had been created in the 1930s by Henry (aka Harry) Beck – of whom, more later.

The Paris Metro map, in contrast, appeared to me to contain no such neatness: looking for all the world as if it had not been designed at all, and merely depicting a slavish representation of the exact trajectory of the wavy, wobbly ways woven by the tracks below (and in some cases, above) ground. It felt as if our French friends would have been affronted by any move towards simplifying the situation. Yes, the surface geography had been removed with the sole exception of the river (as it had in London), but the Seine was shown meandering (as it does) through the city, and the sub-surface kinks in the lines (unnecessary in a diagram – and mostly unnoticed by passengers, even inside the trains) were gratuitously adhered to. Another quirk was the interchange symbols, circular pies with triangular coloured wedges (with Gallic chic, described to me later by one of the re-designers of the Paris map as 'petit Camemberts').

A further conundrum in my infant mind was this: given the chaotic nature of it, why that Paris map *did* contain several schematic aspects. The 'RER' lines (a heavy-rail system that ingeniously links older suburban surface lines via hugely expensive new tunnels under central Paris – think London's recent Elizabeth line) were all squashed onto these maps using highly London-esque diagrammatic forms. On these, the station names were shown not horizontally but skewed ninety degrees. It wasn't as if the French did not know how to draw a straight line or make a diagram – so why not apply that logic to the entire Metro network?

We returned to London and were deposited at the school gates, sticky, tired and just a little more cosmopolitan than before, without having been mislaid by our teachers. (Although I did sneak back onto the Metro on the final day of our trip, unaccompanied, while the others were shoplifting tacky gifts, only to be castigated more strongly than my fellow villains.) I retained my awe of the Paris Metro and my queries about its maps for a lot longer than the plastic Eiffel Towers stayed on the mantlepieces of my friends' parents. Our local libraries had precious little about the London Underground, let alone the Paris Metro, leaving unanswered questions about why they were so different: that clear, well-spaced diagram for London and a much untidier geographic representation for Paris.

My dear grandma, recognising a spark of individuality (and an obsessive personality) when she saw one, ferreted around in an old bureau, which smelt of decomposing paper, and

retrieved a tiny 1920s gem she had held on to. It was a beautifully preserved green-covered pocket London Underground
map that did not resemble the modern ones, but, like its
Parisian cousins, was more geographic – full of wavy lines
and without any hint of the diagrammatic. It was a major
revelation; we hadn't always had the clean lines of Mr Beck.

I was further gripped: the schematic solution was, in my
mind, so obvious that I could not see why it had not been
adopted by our Parisian friends.

The principles behind Beck's design concepts are now
cloned by virtually every transit map on earth (see below) and
Beck's version has been celebrated in several blue plaques,
a play and many exhibitions. The London Underground
diagram has been a BBC *Design Quest* winner (2006),[1] was
celebrated on a postage stamp (2009),[2] stylised for inclusion
in UK passport pages (2017),[3] and used by the Royal Mint on
packaging for coins celebrating 150 years of the Underground
(2013).[4] But how did his winning design come about? How
did it go from there… to here?

2.

Mr Beck became an employee of the main system operator
Underground Electric Railways Company of London (UERL)
– a private company – as a junior draughtsman in their Signal

1   https://www.bbc.co.uk/pressoffice/pressreleases/stories/2006/03_march
/03/design.shtml
2   https://transitmap.net/underground-design-stamps/
3   https://transitmap.net/passport-tube-map/
4   https://coinhunter.co.uk/c/1/

Engineers Office in 1925.[5] The 1920s were a decade of major Underground expansion, driven by senior manager Frank Pick and his favoured architect Charles Holden. Extensions were planned and opened across the city, and most of their architecture was in the modernist contemporary style, which borrowed much from the Bauhaus and other fresh European perspectives. The clean streamlined look came to be known as Art Deco, and it infused buildings, trains, publicity posters and even seat coverings ('moquette'). The only aspect of London's Underground that did not reflect this modernity was the system map.

I had learned that Grandma's green gem was designed in 1925 by an F.H. Stingemore – but unlike Holden's new, sleek stations on the extension to Morden, the map looked distinctly old hat. By chance, the economy was faltering and the UERL was suffering financially; hence, it laid off some staff, including Beck. During his enforced employment exile, he began sketching some thoughts on how to tidy up Stingemore's Underground map as if he 'was using a convex lens ... to present the central area on a larger scale'.[6] In an exercise book, turning the Central London Railway into a horizontal base line, Beck roughly sketched straighter route trajectories, using only one diagonal angle (forty-five degrees) and evening out the spaces between stations. Pleased with the initial result and without any sort of commission, he converted his idea into a 'presentation visual', colouring all the lines and filling in the station names by hand. This he circulated in 1931 to some former UERL colleagues – including Stingemore

5   Caroline Roope, *The History of the London Underground Map*, Pen & Sword (2022), p. 89
6   Ken Garland, *Mr Beck's Underground Map*, Capital Transport (1994), p. 17

– and they encouraged Beck to submit his idea to the publicity department.

To Beck's surprise the concept was initially rejected as 'too revolutionary', but he persisted and re-submitted the idea a year later. With the quasi-nationalisation of London Transport (LT) in the offing (1933), the publicity department changed their minds and called Beck back to inform him they were going to use his design. Beck produced artwork for a printed proof, which he annotated during 1932 and, for the final artwork, swapped the station markers from circular blobs to neater ticks.

When Beck's diagram was first issued early the next year, the front cover proclaimed it as 'A new design for an old map'. Its popularity was soon evident: the initial print run of 750,000 had to be boosted by a second run of 100,000 in February 1933. Beck then worked on large (quad royal) posters for the station walls and retained custodianship of the diagram's evolution for thirty years. Despite many improvements and what some critics have called 'fiddling' with his own design, the map essentially remained the same throughout Beck's tenure. But by the 1960s, the then publicity officer for LT Harold Hutchison decided he could improve on Beck's work and he redrew the diagram from scratch. Hutchison's quirky, angular and unattractive solution was quickly usurped by a much more elegant version created by Paul Garbutt in 1964, and various hands have tinkered, fine-tuned and revised the diagram that is still in use today. In 2006, Beck's name was added to the bottom of the current diagram to acknowledge his initial inspiration for the concept.

3.

While I was still at school and not knowing Beck's story at that stage, the inconsistency I had spotted between the London and Paris transport maps got me doodling my own schematics. In the cavalier mode of a teenager, I concocted rail transport systems for other UK cities that did not have them, drawing or painting each in a Beck-esque diagrammatic style. Meanwhile, my family was about to leave London permanently.

My dear dad, who had willingly followed disused railway lines with me, nevertheless suffered mightily from poor health and he craved a calmer life, away from the arduous schedules of restaurant work in the capital. He applied for three positions far from the city, pledging to the rest of us that he would 'take whichever one offered a job first'. Thus, during the long hot summer of 1976, a thirteen-year-old brought up in London with a fascination for its transport found himself a metaphorical million miles from mass rapid transit: on the perennially prosaic Isle of Wight, in a rural village of a few hundred people.

To my great surprise, I quickly discovered a quirky link with the London Underground: the island was the location of a survivor of the Beeching axe of branch lines, the quaint, short shuttle between Ryde Pier Head and Shanklin. British Rail had decided, in its infinite wisdom, to electrify it but due to there being no other suitable rolling stock, a batch of former London Tube trains, which were about to be retired, were co-opted. Hence the Isle of Wight became the only place in the UK where one could find working Underground trains in service outside the capital. Unsurprisingly, this became yet another object of my childhood fascination, and

I was soon sketching implausible extensions of the existing line (almost to our own front door, of course).

We had learned in Geography lessons about the new city of Milton Keynes being established around this time, so naturally I painted an imagined Tube system map for it (though, to the new citizens' chagrin, the best transport they ever got was a fleet of single-decker buses). In History (taught so turgidly it's a wonder any pupils' interest was ever sparked in anything ancient), we heard about the industrial revolution and the crucial creation of the Liverpool–Manchester railway (1830). At home, using Dad's old road maps (I can still recall the bright yellow covers and idyllic vignettes of the National Benzole company ones, bucolic views of locations long gone), I sketched an imaginary Underground linking Liverpool and Manchester and running out to their suburbs, despite never having been to either great city.

One of my teenage jobs was delivering newspapers and on a chilly Sunday morning I dropped someone's reading material on their path. As I was stuffing it all back together, a page flipped by in the *Sunday Times* Magazine[7] and caught my eye: an unfamiliar diagram of a transit system. Crouching in their porch, I had discovered an illustration of a re-imagination of that peculiar Paris Metro map... as a London Underground-style diagram! I still have the page that I carefully removed from the unwitting recipient's magazine – it inspired me more than it would them, I'm sure, so please forgive me No. 27 Wyatts Lane for your missing page that day. Spotting that image, it was as if a light had been turned on; I was not alone in my funny fascination.

7   *Sunday Times* Magazine, 11 June 1978, p. 79.

Some of the next items added to the repurposed sock drawer (now relocated to the Isle of Wight) were metro and local railway system maps of Glasgow, Liverpool and New York. Each of these seemed to adhere to the general style of the familiar London Tube map. Scotland's only subway was depicted alongside local commuter rail routes radiating from the city under the control of Greater Glasgow Passenger Transport Executive, which had almost slavishly followed the diagrammatic principles adopted by Beck. It was a most pleasing small square card, folded just once (unlike the bi-folded rectangle of London), and not dissimilar to the schematic made for Merseyrail, although Liverpool's used a lesser colour palette.

The much larger fold-out plan of the New York City Subway was the most colourful: its vibrancy exaggerating the vivacity of the faraway location seen only by me, as yet, on TV or in films. I was not generally permitted to stay up and watch programmes like *Kojak* but I craved glimpses of the New York City Subway and longed to one day visit it with the prized and beautiful map, procured for me (like the others from exotic locations) by friends or relatives. It was only later that I learned that the 1975 NYC map in my collection then was a design classic by Massimo Vignelli, whose company (Unimark) had also conceived of a fresh subway wayfinding system in the late 1960s. By the time I got to NYC, the subway map had been radically altered from my prized one.

I was not destined to explore worthy academic subjects, like Maths or the sciences, at university. But at school a few attentive teachers had spotted my design peccadilloes and encouraged me to study graphics. The nearest art college was in Southampton, so my art teachers, Miss Pam Horsley and Mr Jim Mason, encouraged me to produce an illustrated report about the London Underground and a painting in which I re-imagined its famous Tube map. My school's only books on the subject were a dry two-volume academic study by Barker and Robbins[8] (little did I realise then, a seminal work, much sought-after by historians), but they omitted mention of Beck's map. So it would be true to say that at this tender age, while I had learned of his name, I had yet to appreciate Beck's significance, other than having noticed that the Paris and post-1978 New York City transport maps did not adhere to his style. Hence my own attempt at reimaging the London Tube map with a more geographically accurate depiction of the central areas' lines – while adhering to Becks forty-five-degree angles – utterly lacked the finesse and panache of the master. It did, however, curry enough favour with the assessors at art college to grant me a place on their graphic design course, starting in 1980.

4.

Southampton's 1980s art college building was not the finest advert for the doctrine: a drab extension to what had become part of a higher education establishment. It did offer rather fine perspectives onto one of the city's prettier parks and was often bathed in a particularly pleasing light. Contrary to the perception of art as being a subject where free will and self-expression would be encouraged, some of

8   T.C. Barker and Michael Robbins, *A History of London Transport, Volume 1,* George Allen & Unwin (1963); T.C. Barker and Michael Robbins, *A History of London Transport, Volume 2*, George Allen & Unwin (1974)

the vibe here felt less dynamic: the atmosphere resembled an overgrown sixth form and the course itself was geared towards a somewhat regimented processing of students directly into graphic design jobs. Many students desired just that, but I was not inspired by life-drawing nude octogenarians, packaging manufacture or marketing techniques. Photography, lettering, typography, print and media did, however, spark interest, as did an awakening sexuality and a passion for music. These areas overtook my life, diminishing the time I could spare for railways and cartography for some years.

By the early noughties, after various cities and broadcasting jobs, my map collection had expanded to many boxes, and friends and friends of friends were liberally borrowing from the collection – all pre-Google maps. This was becoming problematic: some maps were never returned and others came home to me in a poor state of repair! It was at this point that I had an epiphany, one which would send me racing back to my childhood fascinations.

5.

There is only one thing worse than a book not existing, and that is: having to create it yourself. Sorry to reword the outstanding oratory of Oscar Wilde but this revelation followed the exasperation of:

   a) my beloved map collection being damaged while on loan,

   b) the (then) unavailability of local transport maps online,

   c) the lack of any compendium of such maps, and, most importantly,

d) the realisation that since a book containing all the world's transit maps did not exist, it was necessary to create one.

It was not that I wouldn't relish the achievement of writing a book about metro maps, nor perhaps even conducting the research; it was the understanding that there would be no book unless I could convince the transport operators of each city to allow me to reproduce their precious intellectual property (IP), to be viewed alongside all the others from around the world.

I prepared a dummy of the kind of book I wanted to write. Using colour photocopies (illicitly made on my employer's fancy machine) and even spares from my own collection, I cobbled together about a hundred 'pages' of my compendium of transport maps. Matching the landscape format of Ken Garland's Beck book, each of the major cities was allocated a double-page spread with its current official maps on the right and text (or even earlier examples of that city's maps) on the left.

The book's name was a quandary: the initial working title was an invented portmanteau: 'Metro Maplas'. It was daft and did not survive the mock-up but gave me something to work with. For the cover, I imagined a new diagram showing all the cities featured in the book and linked to each other in the Beck diagrammatic style. Designing this diagram and the other processes took around six months of my spare time but once a mock-up was spray-mounted and stapled together, I contacted the only publisher that I hoped to have a chance with and, to my great surprise, Jim Whiting, the boss of Capital Transport, agreed to meet me, liked the idea and agreed to publish it… with one proviso. I had to get all the copyrights cleared first, starting with Transport for London (TfL).

The ball now in play, two London Transport Museum mates, Mike Ashworth and Mike Walton, introduced me to

David Ellis at TfL: an enthusiastic and affable fellow, who, though famous for his staunch guardianship of the company's IP, was keen on the idea of my book. David also approved of me using TfL's famous typeface, originally designed by Edward Johnston, and the museum gave permission to have their logo on the back cover – each of these concessions a great honour.

To seek permission to use the other official maps (never achieved before in book form, remember), I needed to contact all the transport operating agencies. This was easy enough for the English-speaking countries, and I prepared politely worded begging letters for operators like those of the Glasgow Subway, Tyne & Wear Metro, New York City Subway, Chicago 'L' and Boston 'T', plus other choice British, Canadian and Australian cities. The hiccup came when confronting how to translate my request into other languages. Google Translate did not exist in 2002, so French, German and Spanish friends kindly helped me for the letters to our near-European cities. As I was not at the time acquainted with any Chinese, Greek, Italian, Japanese, Hindi, Portuguese or Russian speakers and had no money to pay for a translator, I used the *Yellow Pages* to find local restaurants specialising in the cuisine of these countries. Then I brazenly walked in to one of each and spoke with the staff about helping me translate my begging letter to the transit operators. Incredibly, many friendly restaurant workers in London's West End agreed to assist and eventually I had the wording for letters to over 100 cities worldwide in over twenty languages.

All I had to do now was find out where to post them! With the internet in its infancy and many transit operators lacking a presence there anyway, some logical thinking was required. The LT Museum came to the rescue again: their superb librarian Caroline Warhurst unveiled *Jane's Urban Transport Systems*, from which I gleaned copious details. The letters in their beautiful languages were duly posted and, as replies trickled back, Mike Ashworth noted I was lacking many historical maps and suggested we visit the International Association of Public Transport, the UITP, in Brussels, to get their backing and search their library.

With funds limited, we flew on a budget airline to Brussels South, which turned out to be over an hour from the city in the former industrial stronghold of Charleroi. But that proved fortuitous as it gave us an opportunity to inspect the local light-rail system before heading to Brussels. In discussion with Mike, I resolved to include Charleroi and all the other smaller metro-like systems in the book but that meant increasing the page count and more translated letters to beg permission to use official maps! The UITP were most generous: their then president, Hans Ratt, agreed to write a short introduction to the book and allow their logo to be used, too, while the library supplied many examples of historical maps that neither Mike nor I had seen previously.

For the cover itself, I suggested to Jim that we could use a traditional desktop globe but with the World Metro Map somehow wrapped around it. I met with Alan Foale (then designer of the London Tube map) who helped turn my scrappy mappy into something more presentable; the similarities to the Beck principles and station markers, line colours and 'feel' of London had spread across the planet! It was used as the frontispiece in full and duly curved onto a photograph of my dad's old desktop globe for the cover. This felt like a perfect marriage of my cartographic heritage.

The book's title was altered from my somewhat idiotic 'Maplas' idea to the more obvious *Metro Maps of the World* and it was due to be published in autumn of 2003. Now, publisher Jim was not known to spend much on promoting his books, sending out review copies or holding launches, but with my media background I cajoled the poor chap to invest in these areas. We printed some small fold-out colour maps of Alan's World Metro Map, at the same size ratio as that used by the official Underground pocket map. These we handed out to delegates at the UITP World Congress in Madrid, where I was pleasantly surprised by how much keen interest they generated (and I even met some of the people I'd previously written to). The LT Museum agreed to host a launch event and Mike Walton, who ran their shop, was so taken with the World Metro Map he had several hundred printed up as a wall poster and his prediction of its popularity proved prescient. Most significantly, the museum shop was preparing their December advertising campaign and wanted to include my book. Their *12 Days of Christmas* campaign that year was a series of oblong card panels to be displayed inside Underground train cars, above the windows: a traditional location for the Beck-style map of the Underground Central area. Imagine my pleasure at seeing the front cover of my book, with Dad's old globe and my world map design, sitting right next door to a card showing the very map that inspired it all!

It was a momentous moment for me.

In the following years, my book was updated several times, translated into other languages, picked up by Penguin Books in New York and revised with new entries under the title

Transit Maps of the World. This in turn was translated into other languages and led to new commissions from Penguin. The World Metro Map took on a life of its own outside the books, too: an American university asked for the artwork to make a huge version of it for their international student's hall. A hotel chain framed hundreds of them to decorate their guestrooms. It even went viral online.

The scope of a schoolboy's obsession had grown into a second career. It allowed me the opportunity to publish multiple books, give lectures around the world and ultimately to meet my boyhood icon, Massimo Vignelli. It was in 2009, while I was working with the planet-sized brain of Peter Lloyd, researching his book on NYC Subway map history, that we met Vignelli. Peter recalls, 'He was in his late seventies and his apartment/studio was immaculate, pristine, systematically organised. A kind of sacred space where the prophet of modernism would welcome all seekers of wisdom and generously give his precious time and insight. Despite being efficient and rigorously organised, it was not cold or sterile but pulsing with creative energy.' Vignelli showed no interest in retiring, having just completed a remix of his hallowed 1970s NYC Subway diagram for *Men's Vogue*. Peter laid out his entire Vignelli collection on the floor of the man's apartment while Massimo ruminated over the demise of his map in 1978.

We each left with a pristine signed copy of the *Vogue* diagram. Peter also encouraged me to take the train to Washington DC to investigate the Library of Congress (LoC) map collection and I duly took an Amtrak south – musing all the while on why America's trains are so slow compared to Europe's. As I was still dirt poor, I had arranged to stay in a private BnB where the host regaled me with a DC 'secret' I must try to gain access to: the Capitol subway system. But first I had an appointment in the LoC itself: the Geography and Map 'room' is a bit of a misnomer – it is a labyrinth of interconnecting rooms. I had researched the catalogue online and pre-booked to look at many items, but was taken into an area which resembled that scene at the end of an *Indiana Jones* movie where there are so many storage chests they seem to go on into infinity. The librarian gave me a drawer set number which purportedly contained 'anything railroad related', adding wryly: 'We've no idea what's in there, it's simply not been catalogued.' Pulling out sheet after sheet, I was astounded by railway maps from across the planet. Wartime Russian and German maps of enemy territories; intricately coloured Japanese maps of mountainous branch lines; nineteenth-century 'bird's-eye' views of railroads into Boston, Philadelphia and Washington streetcar maps...

Hours passed and hundreds of images later I was unceremoniously ejected at closing time. 'Can I see the secret subway first, please,' I begged, but as the LoC was not connected to it I had to run to another building where security regulations did not permit access. This was a shame because the original route between the Capitol building and Russell Senate Office dates to 1909. There are now two more short lines to other House and Senate offices and each tunnel is spotlessly clean, mostly gleaming white and quite unlike any other subway in the world. Though two of the routes are reserved for staff, visitors are permitted on the Russell line, provided they are booked on the Capitol Tour and turn up before five in the evening without a ticket!

6.

None of the plentiful and at times most peculiar antics in my career equate to the great skills of genuine cartographers or the contributions of legions of dedicated people who run, maintain, plan and operate our great transit systems. I think of myself as nothing more than a lay commentator on their achievements and a staunch advocate of increased funding for more and better public transport. Although I tried to put these interests on a more professional footing later in life – embarking on a degree course (gaining a Master of Arts in Railway Studies in 2024) – it is peculiar to reflect that the coming-together of my quirky childhood take on Beck's classic design (tackled again in my forties) had led to my odd doodlings being displayed in the very spaces where the Beck diagram is usually shown. What this signals is that with the good fortune of meeting wonderfully helpful and nurturing people (in my case, those at TfL, the LT Museum, all the transit operators, and publishers, the UITP and the many map collectors and designers), a nerdy geek with niche interests *can* make a little contribution. So, if you're reading this wondering how your own railway story will be told: please do be bold, go with your heart and believe that anything is possible!

Smethwick Rolfe Street station mural symbolising communities working and growing together for the future.

SOMEWHERE AND NOWHERE
VICKI PIPE

**How money, nostalgia and hope have
shaped the UK's railways**

Have you ever stood on a platform and felt the terrifying force of a passing train as it thunders through at speed? It's almost cinematic. Just before it happens, there is a calmness. The rails sing the gentlest of warnings – vibrating at a pitch almost imperceptible as the loco approaches. There is a split-second silence before the shock of the air whips your breath away and forces its way through you. The whole station reverberates. Doors to platform waiting rooms clatter back and forth, the plastic liners of litter bins dance wildly, straining at their miserly constraints. At the exact same moment, the sound erupts. It's not just loud, it's full and wraps around you, jamming all other frequencies so that for those brief moments there is nothing but the screech and drone of metal on metal, as carriage after carriage races by. Then it's gone, as quickly as it came, and the chatter of voices and the trill of birds singing filter back into your ears. The dust settles and recalls you blinking back into reality.

It's these moments that remind me how much of a force of nature the railways are. Not literally, of course, as on the whole nature evolves to the benefit of itself, slowly, carefully, thoughtfully. The speed and dissonance of a passing train is, rather, a reflection of something more unnatural, something that is entirely of human construction; manufactured and re-engineered in a constant cycle for over two hundred years to go further faster, to carry more with economic efficiency. Therefore, perhaps not a force of nature, but a force to overcome nature.

We often forget that at their origins the railways were machines of industry, built for the benefit of goods, not people. Forged with an appetite for expansion that carved deep scars into our geography, took the lives of countless unnamed men and women, deepened the country's reliance on products made by enslaved persons, and fed the profits of the owners of enslaved African people and those involved in the transatlantic slave economy. And yet these dark beginnings, seared deep within the railways' DNA, are perpetually obfuscated. Often described as the Golden Age of Steam: a nostalgic turn of phrase that casts a smokescreen to hide the reality behind celebrated and undeniable feats of construction, the visionary ambitions of daredevil engineers, and the blossoming of regional trades whose craftmanship was hailed the world over.

If you pay close enough attention, you can see physical reminders of this confluence of realities echoing down the centuries. None rings more loudly for me than on the approach to Liverpool Lime Street where, between the modern stations of Broad Green and Wavertree Technology Park, the train trundles two miles though a steep sandstone cutting. Here the track sits at a depth of around twenty-four metres, and from inside the carriage you feel ant-like as your eyes slowly climb the face of the jagged monolith banks of the Olive Mount Cutting that surround you, the top impossibly out of view. Even more sobering in this moment is to know it was sculptured almost entirely by hand with just pickaxes and shovels (and a few blasts of gunpowder to get it started) wielded by navvies whose labour was neither properly acknowledged nor compensated. More sobering still is to understand the drive behind such dangerous and landscape-changing work.

Part of the original Liverpool and Manchester Railway – the world's first steam-powered railway to connect two cities – the route opened in 1830 and careered through peat bogs and hillsides, across valleys, under suburbs and over rivers, cutting journey times between the two cities dramatically. An accolade and moment of pride which is celebrated by all who know even the smallest amount about the Golden Age of Steam. And yet five years earlier, in 1825, at the same time the Stockton and Darlington Railway launched, the proposal to build the Liverpool and Manchester Railway was at first rejected, after heavy scrutiny by both the public and Parliament with questions raised about its true purpose. For whom were these reduced journey times being made?

As outlined in its original petition to the House of Commons on Tuesday 8 February 1825, the members of the committee for the Liverpool and Manchester Railway Company stated,

That it would be of **public utility** [emphasis mine] if a Railway or Railways ... with proper works and conveniences adjoining thereto … were made for the passage of waggons, carts or of other carriages, commencing at or near to the north end of a place called the Prince's Dock, within the parish and borough and town of Liverpool … passing through to several other parishes … terminating at or near to the westerly end of Water-street, in the township of Manchester.[1]

Close examination of this petition, as detailed in Robert Carlson's book about the history of the line, included testimony from those both for and against the railway. The vast majority of whom were not residents of Liverpool and Manchester as potential passengers of the proposed railway, but rather landowners, companies and other manufacturing organisations who would either profit or lose money on the railway's development.

Arguments in favour focused heavily on the improvements to be gained economically by the faster conveyance of imported raw materials to manufacturers, whose products could then more quickly be distributed back to merchants abroad. They painted a clear picture of the failings of the existing road and canal networks, which they said were slow and frequently overwhelmed by the volume of goods that needed conveying. Arguments against the railway were largely from those involved in the operation of the roads and canals, and concentrated on proving vehemently that this was not the case. In fact, some said there was not enough trade to go around, and that the introduction of an additional transport service would essentially put them out of business altogether.

---

1   Robert E. Carlson, *The Liverpool and Manchester Railway Project 1821–1831*, David & Charles Ltd (1969), p. 105

The utility of the railway beyond anything to do with wider commerce and the economy appear, from Carlson's account, to be thinly addressed. Equally unconvincing were the statements aimed at dispelling the undercurrent of suspicion that those who might benefit most of all were simply the shareholders of the Liverpool and Manchester Railway Company themselves. William Huskisson, MP and President of the Board of Trade at the time, asserted that he believed the railway *was* for the public good because its shareholders:

> were the bankers, merchants, traders, and manufacturers of Liverpool and Manchester. They had agreed, that no person should hold more than ten shares each; and if gentlemen, would consider what amount of interest could be realized from so small a number of shares, they would perceive that the profit could not be an object. It was the great interest of the trading community, and not the profits that might be derived from the shares, which mainly actuated those individuals to call for this rail-way.[2]

However, the *trading community* and *railway shareholders* were, in some part, one and the same. Charles Lawrence, the railway committee's chairman, was from a family of merchants who were importing goods into the UK as far back as the eighteenth century, with plantations in Jamaica where enslaved labour produced sugar, rum and molasses.[3] The family of John Moss, the railway committee's deputy chairman, was

2 Liverpool and Manchester Rail-Way Bill – Standing Orders. Volume 12. Debated on Wednesday 2 March 1825 https://hansard.parliament.uk/Commons/1825-03-02/debates/c2b9c7c2-08e8-409d-b84c-1874cdf389cc/LiverpoolAndManchesterRail-WayBill—StandingOrders
3 Charles Lawrence https://www.ucl.ac.uk/lbs/person/view/16128

THE UNTOLD RAILWAY STORIES

deeply involved in transatlantic slavery; John himself inherited businesses and money directly related to this, including a plantation in Demerara with over eight hundred enslaved.[4] Robert Gladstone, the railway committee's director, built his fortune working alongside his brother, John, trading in sugar and cotton in the West Indies. His brother was also chairman of the West India Association in Liverpool, one of a list of organisations who voiced their support of the railway.[5]

Alongside the debate about the railway's intended beneficiaries, there also sat an ambition to be the first to fully realise the potential of the steam locomotive. Was it destined to be the next greatest advancement in engineering, with the UK as its pioneer, or was it an unproved scientific hypothesis waiting to fail? The Stockton and Darlington Railway, which had opened five years previously, the first public steam-powered passenger railway, only carried passengers via a steam engine on its inaugural journey. For the next eight years, including the time the Liverpool and Manchester Railway petition was under debate, its passenger services were hauled by horses. The ability of a steam locomotive to operate reliably and safely was challenged vigorously. Mr Adam, solicitor and legal counsel to the Liverpool and Manchester Railway Company, pleaded on behalf of the *hope* of what the railways could become:

> All I ask you is, not to crush it in its infancy. Let not this country have the disgrace of putting a stop to that which, if cherished, may ultimately prove of the greatest

4   John Moss https://www.ucl.ac.uk/lbs/person/view/7601
5   Robert Gladstone https://www.ucl.ac.uk/lbs/person/view/1648741604/ and John Gladstone https://www.ucl.ac.uk/lbs/person/view/8961

advantage to our trade and commerce; and which, if we do not adopt it, will be adopted by our rivals.[6]

It is clear from the language used by all those supporting the railway that their definition of public utility comes from a purely capitalist perspective. Their ambition for its purpose and what they believed the railways would mean to the UK was intrinsically linked to the improvement of industry and the development of the economy, creating more jobs, which would in turn ultimately – it was argued – benefit the public. This wasn't a library or a school; the idea that a railway would bring any advantage, to use Mr Adam's turn of phrase, at a passenger level beyond getting from A to B was not even a consideration. Technically, as a public as opposed to private railway, people *could* travel along it, but it would be fair to say that what distinguished a passenger service from a goods service at this time was a definition of cargo, with people being just one type of 'goods' to be transported. And yet, despite this, the public flocked to use it and subsequently this and many of the other early steam railways in the UK have become synonymous with the origins of the world's many and great passenger railway networks, even though they were not primarily conceived for passengers at all. An article about the Liverpool and Manchester Railway on the Science and Industry Museum's website describes how a platform wasn't even constructed at Liverpool Road Station (which is, confusingly, in Manchester): 'Instead, passengers had to scale the carriages from ground level.'[7]

---

6   Carlson, p. 131
7   https://www.scienceandindustrymuseum.org.uk/objects-and-stories/making-the-liverpool-and-manchester-railway

This image of passengers scrambling up the side of carriages in order to embark is almost reminiscent of some of my own experiences of boarding a train in the UK today. Who of us hasn't witnessed or even been the leading protagonist in the clamber from one carriage door to another as the beep beep beep beeeep of the automatically closing doors rings out, desperately searching for some sliver of space to secrete into amongst an already overly packed carriage. Announcements cautioning us to 'mind the gap as you step down from the train to the platform edge' as we look down to see a two-foot drop beneath us, though fewer than they used to be, are still regular enough for me to question if the user experience of the railways has progressed as far in the last two hundred years as we might think it has.

I will never forget standing on Lincoln station platform during the height of rush hour with station staff dutifully advising the swelling crowd to fold down all pushchairs and remove any large rucksacks they were carrying, indicating where the best places to stand for boarding the inbound train might be. At first I was slightly bewildered at what felt like an over-instructive direction about how to catch a train, until the train itself trundled into the station comprising just two carriages to accommodate at least four carriages' worth of people. The resigned compliance of the people around me coupled with the pre-emptive actions of the staff indicated that this was not an unusual occurrence.

It makes me wonder if our fellow travellers in times past would recognise the feelings of discombobulation and irritation many people now freely express when asked about their travels along the railways. In preparation for writing this chapter, I spoke to many friends asking them about their own experiences. In every instance it took just a few moments before they were regaling me with anecdote after anecdote about late-running or cancelled services, miscommunicated announcements or confusing ticket types, of disproportionately long connection times, uncomfortably hot or cold carriages, queries about why different trains have different liveries, or the angry-making cost of peak-time travel, when it's impossible to receive even the most basic level of service, i.e. a seat. In all my conversations there was an underlying dose of frustration around how the level of service they received did not match that which they believe they were promised when they paid for a ticket.

'It's a public transport service!' they would exclaim, 'and yet it doesn't seem to serve the public very well.'

I've come to realise that this is perhaps because that was never the intention.

The legacy of the railways past as a capitalist endeavour, this strong cocktail of engineering ambition and economic advancement presented as public benefit, is a tension that two centuries of operation has not managed to resolve. Even temporary nationalisation could not recentre the narrative, failing in part ironically because road transport became, once again, the cheaper, more convenient and reliable alternative. As modern commuters we have to endure the contradictions; We consistently feel the heavy hand of commerce, forced to pay peak-time prices for overcrowded carriages and often delayed services, while simultaneously seeing adverts for leisure travel that provides cheaper journeys along exactly the same route; at the same time, the confusion between different train operating companies running services along the same stretch of track makes it almost impossible to purchase the 'correct' ticket, with the threat of fines disproportionate to the original ticket price enforced for any minor infraction; the government announces new investments in railway upgrades for faster journey times and improved infrastructure; new housing on city outskirts creates hubs around which new stations are designed and built, followed quickly by individual train operating companies cutting existing services or closing ticket offices; announcements that 'the next service is delayed due to availability of train crew'; pending strikes amplifying the discontent of employee conditions and benefits; as reels on social media and billboards inside carriages promote smiling employees who encourage you to join the railway family; you can be delayed on station platforms while watching freight trains thunder passed unimpeded, creating a feeling I can only describe as resigned inevitability.

But wait. What if there could be another side to the railway experience?

What if there was a completely different story, uninfluenced by the past, unencumbered by the burden of the P&L where the most important connections have very little to do with getting from A to B and everything to do with the advantage (thank you, Mr Adam) that the railways can bring to individuals. A railway that could provide public utility through the pure act of travel itself, helping to resolve isolation, providing educational experiences, broadening aspirations, developing links to new communities and improving wellbeing.

What if I were to tell you that this version of the UK's railway does exist. That it is the same railway I've just been describing, not a separate or parallel railway with different stations and different routes operated by different companies. The exact same railway.

Twist.

If you look closely enough you can see physical evidence of this confluence of realities echoing around the network. But this time, don't get distracted by impressive large-scale feats of engineering, or the nostalgia of historically significant stations. Instead, focus on the quietest of places, in some of the most unassuming locations, with platforms so small they barely fit the description; it is here you will find some of the greatest and perspective-changing stories about the network.

Inside the waiting room on platform 1 at Bingley station, West Yorkshire, you will find large-scale British Sign Language boards that illustrate the BSL alphabet and examples of how to sign the names of nearby stations. These were created in collaboration with local school pupils to raise awareness of hidden disabilities and improve inclusivity in the community.

Then there's the old signal box on the platform of St Margarets station in Hertfordshire. Here you can enjoy the curated outdoor gallery, where works from local artists, young people and school groups celebrate various aspects of the area.

Although these are small interventions, they have big heart and even bigger outcomes for local communities. Some benefits even extend beyond the boundaries of the railways themselves.

Like the wildlife garden at Alresford station in Essex, which is home to a giant bug hotel made by a local artist and school children. Interest in the hotel's residents grew until Alresford became one of the first official 'bee-friendly' towns in the country.

And the engineers of these interactions? No, they are not part of the Department for Transport, although they are supported by them. They are not part of Network Rail, although they are supported by them. Nor are they part of any train operating company, although they are supported by them too. Not many people have heard of the Community Rail Network, though likely you have encountered members of the network without even knowing it. Perhaps while waiting for a connecting service, you may have simply dismissed them as just another railway worker tending to flower boxes, installing displays, topping up information leaflets, collecting litter, directing groups of travellers to the next departing service, supporting the young, elderly or those with additional needs to select tickets from the self-service machines. Their members will often slip quietly past you on station platforms, their required high-vis making them highly invisible, notwithstanding the recognition they deserve.

If you ever need a respite travelling along the TransWilts line, take a break at Melksham.

There, opposite the entrance of the station but within the car park, you'll find the Melksham Hub Café. Along with your delicious cuppa and a slice of cake, you will also find a haven for local residents. It's a place where baby and toddler groups meet, and those with additional needs can volunteer to gain work experience and build their confidence – all this supported by the regional Community Rail team.

The affiliations of Community Rail are, like the network and its history, complex. They include a central membership service that provides funding and advocates at national and government level; regionalised partnership organisations who lead on projects in their localities; station adopter groups who take care of individual stations; and hundreds upon hundreds of volunteers. In many ways just like the nineteenth-century railway companies that petitioned vigorously in the *hopes* of what the railways could achieve economically, Community Rail Network embodies the same drive and determination to champion what the railways can achieve for the people and communities they serve.

At the Llandudno station in Conwy, you can join the 'Men's Walks' and 'Menopause Walks' groups, and take part in guided excursions with the North Wales Wildlife Trust.

Nothing brings into sharper relief the contradictions of the UK's railways than encountering the work of Community Rail. They remind us of the railway's responsibility to be a *public utility*, redefining this in a contemporary context, not solely as a transactional relationship between passengers and train companies but rather identifying the railway's collective responsibility to effect change in and for the communities in which it operates. It can do this by providing opportunities to forge friendships, improve health, develop skills, regenerate the environment and offer a space where the most vulnerable can be supported.

In the West Midlands, at Smethwick Rolfe Street station, platform 1, to be precise, you will find a mural created by art students from the nearby college featuring the phrase 'Love thy neighbour' in multiple languages all spoken in the town. In the mural's centre two hands plant seeds, symbolising the community working and growing together for the future.

Now boarding

Over the last two hundred years we have seen the UK's railways transition from the hands of private companies, like that of the Liverpool and Manchester Railway Company, to a nationalised service under British Rail, and then morph into a hybrid model of franchised private companies contracted in the *public interest* by the government. Now in 2025, the latest promise is for all routes to come back under government control to 'provide a railway that works better for both passengers and taxpayers across Great Britain'.[8] The choice of words here, distinguishing passengers and taxpayers, feels concerningly deliberate, to perhaps once again conflate public utility with trade and commerce.

The modern experience of travelling along the UK's railways is an amorphous one. At times convenient, at others joyous, often confusing and always uncertain. The tensions between the legacy of the railways' past, passengers' current expectations, and our aspirations for what the railways could be for communities are in constant flux. Two hundred years is a nostalgic milestone, but what if it was more than this. Let's not think of the railway as two hundred years old, but rather two hundred years young. Mr Adam's passionate plea for the *hope* of what the railways could become is as relevant now as it was then, the ideals behind the railways simply realigned to serve a new purpose, and one that puts passengers, not profits, at its heart.

It's the familiar b-b-b-b-b-b-b-b-eeeeeeeep warning
before the cl-clunk,
clunk,
clunk,
of the door locking mechanisms.

Signalling a green light,
before the high-pitched tenor of the engine,
as the wheels engage in traction

---

8   https://www.gov.uk/government/consultations/a-railway-fit-for-britains-future

and the gentle and involuntary jolt in your stomach
as your go from stationary, to motion.

That sway of the carriage.
As the picture framed image of the station platform
with its peeling Victorian ironwork roof
– once magnificent you assume –
but now faded with age,
pigeon droppings,
and the grime of industry,
turns into a rolling motion picture in your window.

You see alighted trains at adjacent platforms.
Refocus on hi-vis adorned station staff
platform paddles nonchalantly tucked into back pockets,
reaching for devices to check the departure time of the next service,
or perhaps a cheeky look at a text.

Speed increases.

The scene shifts again
to the backs of terrace houses.
You wonder how disruptive the trains must be at night
but conclude, there must be some sort of rule that says
trains must stop at a reasonable hour.

Though reasonable on the railways you know,
is always relative.
You can attempt to delay repay,
but the value of your delay is not always viewed
as valuable to the company.

It's the clammy recycled air,
produced by the modernised rolling stock.

The back of seat trays,
with coagulated crumbs
of many a hastily eaten M&S sandwich
and late night boxes of fried chicken
congealed together by the inadvertent drip
drip, drip from now lukewarm teas.

Drinking from inexplicably resilient take-away cups.
Or maybe gins in tins,
three for two, or something for one
you hurriedly rush back from the checkout to grab.
A bargain you didn't know you wanted
from your ritualistic visit to the station convenience store.

Inconveniently stocked,
with either shelves bulging else
everyone dancing around a lonely packet of falafel.
Where floor space is prioritised for stock, not people.
Your pointless apologies for invading others personal space
lost in the back of your throat.

But safe from all this now in your passenger seat.

Your gaze is diverted to another window.
To check on updates, scroll mindlessly,
read messages you could respond to
but will later justify as 'ah, sorry I was travelling'.

It's the permission to escape the world.
To be absent while in motion
but to always know exactly where you are.

The next stop on this service will be
Watford Junction,
Rainhill,
Salhouse,
Airdrie,
Newport,
Sholing,
Langley Green,
Mountain Ash,
Par,
Rhyl,
Roy Bridge,
Finsbury Park.

It's the contradictions.
The past in the present.
The expense and the value.
The public and the private.
The solo and the collective experience.

To be heading somewhere,
while in the moment being nowhere.

You sense the anticipation to see friends.
The nervous excitement for the first day of a new job.
The laughter of the toddler across the aisle,
held steady by loving arms

enthralled by the motion,
wide-eyed as the blur of a southbound service
whizzes
by.

You see feverous conviviality ahead a seminal cup final,
contrast the earnest studiousness of the avid reader.
Hungrily turning page after page
oblivious to disrupted service announcements,
of disgruntled travellers,
or unexpected delays at red signals.

You share wry smiles, sighs, and sympathise
With those you never knew, will never know,
but who share with you in this moment,
between somewhere and nowhere,
the fluctuating fortunes of a journey
two hundred years in the making.

Nairobi Railway Museum.

A TIMETABLE FOR GHOSTS
YVONNE A. OWUOR

(in ten parts)

Prelude

Morning. The new-old Nairobi (Central) Railway Station. Old, because the infrastructure of its history pulses just beneath the anodyne skin of modernity, a modernity that has gutted its flesh, but not its soul. Suddenly 'Central' because an interloper, Syokimau, has entered the story. The former drop-off zone, once an elegant theatre of arrivals and farewells, has become a sprawling park for *matatus*, our passenger-carrying metallic vibes, metaphors of a city moving on the edge with swagger. A nation's archive and mood menu on wheels. Ironically, perhaps inevitably, some of these *matatus* now ferry passengers to the 'new-generation' station in Syokimau. Much has been made of the 'ultra-modern ticketing system'. (Scoffing inwardly, I am forced to confront the emergence of my inner Luddite.) The UMTS resembles every other UMTS in the world: clean, efficient and digital. No distinguishing detail, no spirit of place. No evidence of a designer who stopped to ask, even once, what Nairobi might look like when turned into code.

Where once ten thousand footsteps echoed now lingers only the straggling tread of those under-using the Nairobi Commuter Rail, big on hope, indifferent in delivery. The vast concourse feels hollow, oversized for its role. Whose really bad idea was this 'commuter rail' thing? (Maybe it is also my curmudgeonly resentment, the feeling of going to an old, beloved home of memories and finding a wasteland.)

Memories converge. Memories of old hastes. Nostalgia for wheel-less luggage, for breathless latecomers, for the orchestrated chaos of train-event management, a world once alive, now in shadow. Vestiges linger. Like the old railway restaurant, its bones still standing, but the soul? Gone. No longer the dense art of yearning and hunger. No longer the pulse of rail-culture bustle.

Of time loudly ticking. (We used to hear the minute hand fall into its quarterly grooves.) Announcements crackle arrivals, departures. Of the steel and steam and smoke, and grandeur of a cult of movement, of transport by these big, brash trains. The station looks fragile, now. Frail. The station now, fragile, frail. Ghost of its own legend. When I was a child, it might have been the gates of hell, or the threshold of purgatory itself. Such grandeur, such awesomeness. Gone are the bell-like 'du-doong' that preceded the dulcet Kiswahili, the English: 'Mabibi na mabwana... Ladies and gentlemen...' And then we would hear some train-based update, and it mattered to each one of us.

Now the Nairobi (Central) Railway Station stands, with its weather-worn age on its multi-triangular head, in the belly of the loudest part of the city. Its trunk is dotted with the new Kenya Railway Authority-branded colour: yellows, browns, the rather toothless ineffable shades that are neither this nor that.

It is not grand, this station, but it is haunted.

By souls in transit, by longing and losses. By the million, million older feet that have hurried toward trains about to leave. The smell of diesel. The clang of gates. The crackle of intercoms. The prayers muttered while waiting. The final hugs. The missed goodbyes. It breathes. It waits. It holds. The station still does not sleep. Wearing its tame colours with unease, it listens. It dreams. Footsteps. Trains. Farewells. Regret. Loss.

Memories.

One day, a teenager still, I chased the night train to Mombasa at that station, the train in cream and dark purple livery, its engines already rumbling, I arrived late, heart thudding; just past the entrance, my dark blue suitcase burst open, spilling my weekend life across the platform, shirts, books, dreams scattering into the dust, the tubby station master barked at me, but with a flick of his wrist (magic!), he held the train back, the engine huffing and puffing, restless to leave, a porter in a faded uniform knelt beside me, helping gather the mess, hands swift, a woman I could not see called my name, but I did not turn, I did not answer, I rushed into the wrong carriage, doors clanged shut behind me, the station master blew his whistle, dropped the flag, the train surged forward, and I, dragging the re-latched suitcase behind me, found my way to my second-class cabin, blue seats below, top sleeper berth, and sat in the dark, breathing out the cold Nairobi air of sweet departure.

Memory.

Last year's silhouettes.
　　Echoes.
　　Those announcements in warmed-up tones:
　　'Train to Kisumu now departing Platform Two. Final call.'
　　'*Mwito wa mwisho. Gari la Kisumu linaondoka sasa kutoka jukwaa la plli.*'
　　Remember how we used to run? Remember the sprint of bodies towards already-moving doors. Voices of the nations on the platform, the cadences of strangers: those tourists, destination-seekers.
　　And then, the delicious grand exhale of the engine.
　　The station master's choreography.

Departure.
　　Echoes.
　　That feeling.

I: The Invention of Place

The station and its city carved itself out of the ambition of feral men, imperial, yes, colonial, indeed, but also wild and mad, who determined to build the impossible railway, their efforts the subject of mockery, until it wasn't. A camp called Nyrobi (with a 'y') grew around the project in the swamplands of *enkare e nairopi*, place of still waters, a transit zone that would explode into a regional capital city. The metre-gauge railway was their dream of world-imagining order, of rule, of regulating nature, of the display of industrial mettle – steel smoke, power and of movement.

'It is not an uncommon thing for a line to open up a country,' mused Sir Charles Eliot, then Commissioner of 'British East Africa', in 1903, 'but this line literally created a country.'[1]

The railway was on its way to Uganda, but then Kenya happened, Kenya with its main-character energy, which it has kept to date, much to the annoyance of its neighbours. Kenya already existed, true, by so many names and senses, relationship and notions, other worlds and rhythms of being. But the railway did take a thousand distinct, uneven and colourful threads and stitched them into a single flag. A train to rule them all. A track to bind them. And so the land entered into new place-names: *Nairobi* (the swamp of cold water) became a capital. *Voi* (the snake river) turned into a depot. The coast marked its time by the arrival of steam. The 'iron snake' rewrote Kenya's next-world story, its culture in movement, and industry. It inserted itself into this Kenya's voice migration, desire, ambition, the meaning of its departures and arrivals.

The platforms that would generate so much milling and noise and anxiety are quite empty now. Emptied.

A pity.

Time here is as mercurial as this city.

---

1   Charles Eliot, *The East Africa Protectorate*, Edward Arnold (1905), p. 208

Disloyal.

Promiscuous.

Transitory – like its trains.

Won't keep familiar worlds behind a plexiglass window, so we… I… can peer in and think…

There. That was a beginning. Here, that was how the ending started…

Daydreaming in the train station that birthed a city:

Of the uniformed curators of nation-in-reformation human arrivals and departures.

Of friends and aunties who arrived and left by train.

And whistling porters. And mountains of luggage, and quiet battles over the weighing machine, a contestation of extra kilos that cost.

And wooden benches for all the waiting people…

And that restaurant where they sold eggs, masala tea, chapati and scrambled eggs…

And…

No train sound.

No announcements. No conductor.

No river of passengers.

(Just a trickle.)

No *ritual*.

To those who remember, this absence stinks of sadness.

And betrayal.

How dare the city change its face so rudely and without warning. How dare it succumb to the visionless leaders who, like feral wild dogs, hunt only for profit and convenience and function, even if it means stomping all over the city's soul, with no emotion, no nostalgia, no ache for the strangeness that formed, not only a swamp into a capital city, but also imagined itself into a nation that had been lurking, anyway, in-between the world.

II: The First Stone, The Last Ghost, Those Trains

Threshold station. Threshold nation. Loyal only to itself, a reluctant harbourer of life's exiles, mystic home of four big rivers and a thousand tributaries with mirrors above, and ghosts beneath.

The station was built in 1899. It blossomed in multiple directions. It hosted the first all-stone building, constructed in 1903 – a Catholic church that was the first to leave the place to occupy its own heart as the grandiose Cathedral Basilica of the Holy Family. Do the passing city's people know this? About this station at the Edge of Time, the capsule of the Other Kenya Genesis myth?

The new 'fabricated in China' trains are streamlined and polite. They do not belch, and bellow, and chukuchuku, as if they alone were the signifiers of the passage of time. The large central clock that used to determine whether passengers should panic or not is long gone. The accoutrement of 'railway' have given way to the digital realm, or breeps and burps, and screens and lights that flicker on and off all on their own.

Vestiges of railway pasts. Floorboards. Benches. Brass plaque of the East African Railways and Harbours Corporation.

The tangible, solid things now fading into a time, still present but consigned to stoic stillnesses: an old gate. Railway fences. Resonances of goodbye. All the entry-departure-waiting symbols. Portals of longing.

There are layers to this place, as if liminal realms have become entangled. At times, silence is sound here. The echo of a generation departing for Makerere, or exile, or war, or love. Of the millions of the comings and goings of Kenya's mobile students. Reverberations of loud and joyful noises of families meeting, of children returning from boarding schools, of the solemn carrying of coffins to final burial grounds.

Of life.

Of life. Of all the goods and hopes that did not arrive in time.

Architectures of Memory in stone.

A station guard sent me to the Nairobi Railway Museum, where remnants of the old railway world are kept, artefacts stripped from another era to make way for the half-hearted new. Inside the building, the corridors still smell of oil and distant travel. There are benches plucked from the Nairobi Railway Station, worn smooth by decades of sitting, waiting, hoping, resting. Windows that once framed hopeful or wondering travelling faces are now assigned museum-approved labels. The infrastructure of that time left behind: signal towers, rusting tracks, faltering mechanisms. Jutting levers. The present semaphores speak in a new dialect, one born of this age, but the old... they pause mid-sentence, these giant life and gatekeepers in metal. There used to be a two-headed board, indicating platforms and the times of arrivals and departures. Souls would gather beneath it, sharing the same frowning disquiet, an expression mirrored today by those who stare at airport boards, seeking the doorways where appointments at a destination become departure.

Does the old station remember when it received the things of life the trains deposited? Or the silver stolen from dining cars, and the first-class, second-class, third-class tickets, and carriages and the souls that rode third-class. The four-chime gong, a xylophone, that the dining-car steward used to summon people to meals, the fine cutlery, embroidered blankets and sheets for the sleepers.

Wandering.
  Wondering.

Here, with the ghosts of the epochs, one finds another of the threads of the making of Kenya. Another identity marker: we are also a railway-formed people. Yes, that too. The land that adjusted itself to fit the railway. Nairobi: mercurial harbourer of life's exiles, loyal only to itself. A trickster city, Nairobi, yard, campsite, a stopover, a crossing, not intended as a destination, existed before it became what gave it its Kenya shape. A liminal portal, the hub of in-between. Home of four big rivers – Nairobi, Mbagathi, Mathare and Ruiru – with a thousand tributaries, with mirrors above and ghost flows beneath. The waters once flowed clean and certain. *Enkare e nairopi*, place of still waters. Guaranteed pastures whenever drought season arrived. Into such a place did a railway arrive and change its plans. Nairobi Railway Station happened. As did the city that would grow into East and Central Africa's

THE UNTOLD RAILWAY STORIES

pre-eminent meeting place. Within the hall of dreams, time takes a break from a now-overloud city in a hurry that races past without stopping to look at this site of mirrors.

It has been announced that everything will be remade. A consortium will build a Nairobi Railway City within a city. A hub, modern and vast. Multi-modal, they say. Mostly digital, they say. Whatever that means, I think. It should matter. It should mean. It might salvage the suffocating of the past. Debris from cynical modernities, politicians' projects designed to fail: a ten-metre abandoned Nairobi Metropolitan Rail Service. Infrastructure of nothing abutting greatness with no shame. The recent place-makers operate without philosophy. No sense of adventure, or the immensities of worlds to create bridges into. It shows: their mediocre desires. In their version of 'railway'.

Storm clouds gather over the city, June's sky bruised with memory.

I find myself staring at the ghosts, of grand locomotives, the steel beasts that once carried the pulse of nations. Magnificent. Splendid. Awe in superlatives. When I was a child and heard the train, and glimpsed from childhood windows, those sounds that accompanied the journeys: whistle, bells, roar, puff-puff, train-on-rails echoes, I knew a Great Thing, a mythical presence roamed the land. They were so untouchable then, like visiting meteors. But now I look behind the veil at the slayed and stilled monsters: the operation cabins have no screens. They only have pressure gauges, brass handles and the machinery of real power. And in this moment, it becomes clear: whatever else is said, before the best of human dreaming was outsourced to anaemic corporations with eyes fixed solely on the bottom line, there existed men and women who dreamed in lifetimes, who were able to put their bodies into five-hundred kilometre wrestling matches with metal, energy, time, movement and desire. Who deferred gratification, laboured through the now and delighted in the legacy their battles wrought. 'Men of steel' is no anachronism. There was no distance then from the fires of world-making, from the forging of futures with all the jutting, lurching steel their hands could muster.

'The journey is the destination.' Who said that in English? It is already a Swahili philosophy that undergirds the very essence of Safari and Msafiri.

The place now.

The station.

Limbo.

The Nairobi Railway Museum.

Purgatory.

The old locomotives, still grand, still shining, look impatient, as if they are done with enforced stillness. They were born to move. They refuse to rot. Yellow, deep ochre paint peeling, stilled clocks, passing crowds indifferent to what this world was, is. Thresholdverse. The long-stop headquarters for in-between people, those that never quite reach their destinations, who still wait for their train to be called.

(That grand train producers revived to insert into Robert Redford's *Out of Africa* is also here.)

And me, us, those whose childhoods are woven with this place of myth and happens, of events that formed an entire nation. A lexicon of encounter, of movement, of adventure and expectation:

Reru. Gari. Reli.

III: Origin Story: A Thread... or Two...

To those who remember, it is easy to weep for what once was. In the Railway Museum, the waiting hall for retired trains, you find traces of the story: from the Lunatic Express, the massive construction project, imperial folly that ushered in waves of new migrants into the Kenya-making project, to the Uganda Railway, for a Kenya that was a mere passageway to the coveted pearl, Uganda. Then the East African Railways and Harbours Corporation, which vanished like a phantom with the collapse of the East African Community in 1977, undone by the wasteful quarrels of ego-driven men. After that: Kenya Railways, a name that has endured, though the thing itself has lived an unsettled, spectral life, subject to whims that birth other train dreams from its ever-murky shadow.

The city, this Nairobi, races past this and other places, the old railway headquarters that started life as the site of the city's first Catholic church. This city, that remembers but chooses to perform forgetting, dashes past its abandoned umbilical cord, now destined, it is rumoured, for a makeover.

Nairobi's marabou storks preside over blackened grevillea trees. They look quizzically at us, my brother and I, and our city.

IV: On Departures and Other Forms of Arrival

How many departures make one arrival? There is a cavernous silence, even amidst the unhaste that is the Nairobi Commuter Rail system. Even I can see that whatever it was meant to be does not flow with ease. I think the station is sulking. It had prepared itself for Charon: stoic, solemn ferryman of eternal thresholds. Not this wan substitute. As though the underworld had sent interns. Everything feels like it is in a state of waiting here. A hesitation held between the closing of one door and the opening of another. *Ma*, they would call it in Japan – the pause between notes, the emptiness that gives shape to future sound. Something hangs on the verge of beginnings and endings.

Thinking about transitions.

When do arrivals and departures end?

There were things we never questioned as children: the voice behind the announcements. The man or woman who spoke into the station's bustle and pauses. We might have sensed their humanity in passing, but mostly, we imagined them as part of the railway's natural order, as constant as the wind, the sun, the stars. They were there. They were mythic beings, making mythic sounds.

Like prayers. Only these prayers were answered, immediately.

'The Kisumu train will arrive in ten minutes.'

And it did.

'Gari la Mombasa linaondoka saa moja jioni.'

The Mombasa train is leaving at 7 p.m.

And it did.

Close your eyes, and you will see it again. A suitcase hauled into the cabin. A hand pressed against glass, a head leaning out through the open frame for one last wave. You think of Voi, of chugging past Tsavo, where you will reach for the window with the sign in red 'Danger: do not lean out of the window', and look for lions, antelope, hyenas, as if they, too, are passengers of memory.

Behind you, the memory of the train leaving the station, the station master's *'Safari njema'*, and then the city's taillights floating down Waiyaki Way, trailing like fireflies, as the train heads east or west across the spine of Kenya. Memory, too, blinks red, then vanishes into the dark. You feel the weight of all you once carried, lodged in the cavernous ache of the heart.

These spaces of transition… Nairobi Railway Station… still hold worlds.

They are not inert objects.

How could they be?

In my culture, the Luo, we say: the stones remember everything.

And if the walls remember, they must project something back to us.

There is a character of grief, shaped by the act of watching history leave.

Dissipate.

Disappear.

Dissolve.

True.

There are trains departing elsewhere.

Even from here.

To Syokimau.

In Syokimau, the 'Terminus', the vibe, the feel, the age…

It is not the same.

It will never be the same.

The aesthetics of journeying are particular. Not all travel-time folds itself into a shared glance of would-be-lovers, the strut of the station master, the sweat of many hellos and goodbyes. Some stay apart, aloof and self-regarding. All those screens. All those machines that simulate talking back to you. All the buttons we, the passengers, are compelled to depress.

All the letting go of the way things were.

I think.

I did not think that I would have to let a train station go.

Say goodbye. Universes are not supposed to die, are they?

V: In the Waiting Hall of Forgetting

I think.

We mistook movement for meaning. The bigger, shinier, more digital and faster the movement, the greater the meaning?

As. If.

The not-Nairobi terminus:

We have conjured new machines to outrun time, but we travel twenty extra kilometres to get to a drop-off point inside a drop-off point where dogs and men will scan you and your goods, and deny you your libations, your home-cooked meals, and the rest between stations, of calling out to the nations along the rail route, of arriving properly in the middle of the city, and not just at its edges.

There will be no slow unfurling of presence, or awakening to arrivals, of sometimes the train shuddering to a stop in the middle of the journey, coughing, stuttering, giving up. Waiting for parts to arrive, parts that will shift the journey onward. In-between the apologies of the captain and crew, we will wait, good-natured. No point in scolding a train. Strangers would gather and play cards, and joke, and shiver together at the cries of other bewildered night beings.

In the new age, shadows are lit into extinction. Everything now is so bright, so light, glass and fibreglass. Everything is scheduled to precision.

Streamlined longings.

They say this is 'progress'.

This SGR: standard-gauge railway. (Is that even a name? Where is its anointing? Where is its essence? What is its thing-ness?) The uncomplicated order delivered in three languages in Syokimau. (Mandarin has now entered the conversation. And Nairobi, as embarkation node, need not apply. For now.)

VI: Nairobi Terminus, Syokimau: A Platform for the Future People

They built this station like an airport, clean lines, glass light, a single great hall to bind us in efficiency for the Madaraka Express. Here, the future hums in air-conditioning and scrolling LED screens, unwrinkled by history. There are no wooden benches worn by waiting mothers, no scent of roasted maize or *mandazi* or boiled eggs through open windows. This is the inheritance: chrome dreams and ticketless gates. Syokimau gleams because it thinks it has forgotten everything beautifully. Twenty kilometres south of the city, away from the city, a new rhythm. Echoes of the emergence of a different empire, still not quite ours. The Great Kenya investment. No hauntings, yet. No ghosts. None of the old farewells baked into the platforms.

Speed.

Order.

Frequency.

No breathing space, not really.

Glass walls. Escalators. Digital announcements. Luggage checks.

A station for all nations, they say.

Doesn't feel the same, though.

Not threshold-y enough. The sounds, all the announcements, are so well-regulated, no echoes, not really. Not Nairobi enough. What I mean is… no, let me explain nothing. Nairobi, headquarters of Threshold realms… the mercurial transit lounge… she will bite back.

VII: Ode to an Ending

Stations are for leaving.

As I must, before the Nairobi Commuter Rail authorities discover my trespassing, the stained tears of drying memories on an older bench in an eternal city.

Hawks. Marabou storks. Wind.

Wheels on tracks.

What was.

What might still be.

These haunted (un)silences.

I think.

All departures are the first departure

(don't know about arrivals, though).

To stand at this railway station is to stand inside a wound. Open. Waiting. Not bleeding, not intending to heal. (What have they done to you, station of my childhood?)

Like transit lounges invested in a thousand, thousand elsewheres.

And Nairobi, heart, the nerve, where the lines converged. A planned happenstance, a rail depot turned capital. Nomadic nations ever-green meeting zone. British migrant blueprints on African earth.

Imperial etchings in landscape.

She is still beautiful, this old Nairobi Railway Station. Watching the city like an unvisited elder, dressed in the robes of an age that is too short-memoried for her. Her own stories of time are sealed in unwanted memory vaults. She is not a palimpsest to be written over. She is a palimpsest, to be written over. One day, someone of the new generation will think to ask how she matters, or means, or knows how to spell the future: *mustakabali*; *hatima* (more like fate; more like destiny).

(Add this: 未来, pronounced *wèilái* – but not to her, the Nairobi (Central) Railway Station. She will toss 未来 into a commuter train and send it off to Syokimau.)

I think.

We, Kenya, ought to have chosen differently. We should have built closer to this, the first Nairobi station, and let our next railway mythos rise beside her. A legend cast in our own idiom to feed the future with the fire of deep memory. So that tomorrow's trains might move not just with speed, but also with spirit, bearing the rhythm of our becoming.

VIII: Have You Ever Arrived at a Train Station after the Trains Have Stopped?

That stillness.
No footfall. No calls.
The creak of sweating walls.
And the murmurs of the station speaking to itself.
Ghosts wait here.
Not loudly.
Not in terror.
Just politely. As if queuing.
For the train that was meant to arrive in 1952.
Or 1982.

Or last Tuesday.

A porter still folds an invisible ticket in his hand.

A mother presses a child to her chest and counts the stars.

A man with a small leather bag asks a stranger if the train arriving is the Kisumu train. Me, I met the trains that had stopped at the Nairobi Railway Museum. They asked me: 'What's the time today?'

IX: The New Line

To the south, Syokimau.

All glass and angles.

A cathedral of speed and function.

And yet –

a yearning, too.

Built by others. Funded by us in secret ways, undisclosed, and the newcomers. This, too, is an implant. Tracks laid now by opaque trade agreement that have transferred the costs of new follies to 'We, the People'.

Debt-laden destinations.

(At the end of one line, the bust of the three-jewelled eunuch, the great commander of the ghost fleet, emissary to the nations of the world, including the East African region, in the age before railways, Admiral Zheng He. I think: He is the captain now.)

New station lives.

New station memories

(not mine, not yet).

Like the making of another layer to the country of layers,

An other Kenya.

Children run in the echo of concrete halls.
A station haunted by possibility.
The first station was *'pole, pole'*.
This one is *'chap, chap'*.
Nairobi remembers.
Syokimau aspires.

X: *Mji wa Kumbukumbu* (City of Memory)

'Even the tracks dream backward': who sang this? I don't recall.
 I know, though...
 That time is a spiral, a curve, an echo.
 Departure and arrival can happen at the same time to the same person
 A nation can be sewn together by a train's timetable.

A station is a cathedral.
 Of human movement. Of absence.
 Of waiting.
 Of yearning.
 Of hoping.
 A station can teach you how to say, 'I am going home.'
 Each train that leaves is a prayer,
 An exhalation of breath flung into the open horizons of closet dreams:
 To return with a name the family will finally honour.
 To find the one who stayed behind.
 To step off into a city where no one knows your history.
 To earn enough to pay the dowry, buy your mother's silence for the clinic.

To rewrite your history in a town without your father's shadow.

To come back lighter, or not at all.

To be forgiven.

To disappear, but like mist.

To arrive and be named *my beloved*.

Each train that returns is also a ghost. It is no longer the same thing that departed, but what it became in motion, shaped by the long exhale of steel and shadow, transfigured by the heart-world of its passengers. When it pulls back into the Nairobi (Central) Railway Station or is carried into the yard of the Nairobi Railway Museum, it has become something else entirely.

Not a machine, but all memory, folded into time and steel.

The Nairobi Railway Station, or its husk and resident ghosts and resident cats, is still there.

Beyond the guards.

Beyond the man sweeping the entrance.

Past the man reading a newspaper beneath the shadows of the past.

The ironwork holds.

Although the signs are fading.

And a distant train chugs.

Chukuchukuchuku.

Today.

Echoes of Anglo-Kenyan, country-singing Roger Whittaker's singing 'The Good Old E. A. R. & H.'...

Railway songs with human rhythms, paced like a heart in travel. Sigh, breath, longing. Songs that keep you company.

(Yeah, yeah, we do need new railway songs, but they would probably come with synth baselines, holograms and laser zaps.)

There is a silence that follows the train. Not the kind that comes before. That silence is still full of expectation. I am thinking of the silence of the aftermaths.

It is the character of the silence here at the old Nairobi Railway Station.

Familiar. Close.

The lover who is fading in front of your eyes.

The losses.

One by one:

Silver cutlery. Clocks. Crockery.

Once passed down from hand to hand in the dining cars, remnants of human civility, fragile rituals moving through landscapes torn open for tracks.

Gone now.

Lifted. Looted.

By some self-declared inheritors of futurity,

those who came claiming to modernise

and left with what time had asked us to keep.

Crowded silences here.

Sad.

Regret:

I wish.

We had the kind of curators of Nairobi's existence who had the foresight, who understood how to carry the spirit of a city, a world across time.

Where was the old train announcement room, I ask a guard at the ultra-modern ticketing system.

Are you travelling?
No.
Are you supposed to be here?
No.
But I came because I was looking for lost memory.
'Go,' he says, 'to the Railway Museum.'

I think.

I heard the crackle and someone clears their throat to speak.

What followed was...
    Silence.

Silence... it lingers in the waiting hall,
    where benches remember the weight of bodies
    and sheltered the waiting,
    including those who had gathered to offload the coffin
    from the cargo carriage.

Silence
    pools in the corners
    where lovers stood, delaying their goodbyes
    in case it was an ending.

Silence is
    The sound of a question,
    unanswered for a century,
    circling the old ticket booth like a moth.

One day, I hope our nation and its leaders and dealers will assemble to collect and collate its memory pieces.

    and

    return them to their places:

    a fork in a drawer,

    a ledger on the shelf,

    a station clock reset to the hour the last true journey left.

The station knows

    all these sounds.

As the night settles over this city, the outlines sharpen. My city's restless people are on the move. The station becomes silhouette. Angular. Still. Its lights flicker on, dim guardians of a vanishing day and era. Shadows stretch along the tracks, soft and long. The city takes over now, swamps old dreams as it races into its future. No closing ceremony for this station – there used to be one – a flag ceremony. The national flag lowered, and, if we were nearby, we would all halt our passage.

    The city inhales, exhales.

    The marabou sentinels are nesting.

    The old station listens, again.

    It listens.

    Again.

Celebratory beers on the Doğu Express as it leaves Ankara
for Erzurum in the east.

THE BAKU BARNSTORMER
JACK CURTIS

Two weeks, 6,000km, fifteen trains, ten countries. This is the Baku Barnstormer. Our attempt to get all the way from the UK to Baku in Azerbaijan by train, without flying, to attend COP29, the biggest annual climate-change conference. I set out on this journey with my two friends and business partners Jacques and Jess – the two Jacks and a Jess.

Our intention for undertaking such a long and adventurous train journey was to see if we could make the sustainable option the fun option and, hopefully, convince a few more people to take the train instead of the plane every once in a while. This was to be a no-holds-barred look at the vagaries of cross-border train travel.

It wasn't about never flying; it was about encouraging people to do what they can when they can. To put it another way, it was about promoting imperfect environmentalism.

We wanted to show that the train is also an option and that when you choose the train, it's better environmentally and, frankly, it's often a lot more fun.

There was just one problem. The land border to Azerbaijan

was shut. Very shut – the only way you could enter Azerbaijan was by air. The sleeper train we would need to take from Tbilisi in Georgia to Baku lay dormant.

As we planned our journey, the closed land border seemed – or maybe we just hoped – like a problem that would go away. It was supposed to be temporary, a Covid-era measure, but somehow it had become the status quo. Surely, we thought, as hosts of the climate conference, they would open the borders? As the trip creeped closer, and the government of Azerbaijan continued to extend the closure, it started to seem like a snag that might ruin everything.

There was still some hope. In theory, the Azerbaijan government could grant special permission for us to enter via land. We contacted the embassy in London and formally submitted our request. Then we waited. And we waited.

Our start date arrived. Every part of the schedule meticulously planned bar that last border crossing. We decided to roll the dice and depart as planned, with our spirits high, but without a response from the Azerbaijan government.

London – 28 October 2024

We set out from the Eurostar terminal in St Pancras Station, London, on an 8 a.m. train to Paris. It was where many similar journeys had begun for the three of us and it unfolded in much the same way. A smooth journey to Gare du Nord and a brief spot of lunch in Paris before grabbing our onward train. This time from Gare de l'Est station for a three-hour train to Stuttgart in Germany, which is where we'd catch the first of four sleeper trains.

Europe seems perfect for the sleeper train and so it's fitting that there's a renaissance underway. A map of European sleeper routes resembles a somewhat orderly bowl of spaghetti, with a rigid tangle of routes heading to all corners of the continent and beyond. These maps barely recognise the borders that mark the constituent countries and offer up a vision of frictionless travel.

This only partially resembles the reality. Throughout our trip, it felt like our cross-border journey was rubbing up, ever so slightly, against the zeitgeist, with countries looking to close rather than open their borders.

We passed through several countries where politicians advocating for closed borders are in the ascendent, and border crossings came to define our impressions of many moments of the trip. Sometimes, they passed without note and at other times they'd upend everything, but we'll get to those.

For now, it was all about the moments that awaited us on the trip. Our previous train journeys had taken us to the south of Spain, Italy, Greece and even Tunisia (admittedly that one needed a ferry too). Each of these trips led us to unexpected places along the way that we never would have visited otherwise, which is why train journeys like this one are so special.

On the Baku Barnstormer those unexpected moments would range from visiting Dracula's hometown and spontaneous dancing in a Bucharest beer hall (for Jess, not for Jacques or me), to botanical gardens in the coastal city of Batumi in Georgia and finding ourselves in a suburban Istanbul hammam struggling with the etiquette, all the way to being presented with flowers by a police officer in western Azerbaijan (sadly, again for Jess, not Jacques or myself).

Stuttgart, Germany – 28 October 2024

We stand on the platform of Stuttgart's Hauptbahnhof station clutching our sustenance for the night ahead – warm pretzels, salads and cereal bars. Our destination is Budapest in Hungary, and we are due to arrive at 11 a.m. the next morning.

We have booked a sleeper cabin. This is important. There is a hierarchy of ticket types on sleeper trains, which can be made more complicated by the hodgepodge of booking websites and Google Translate's varying levels of cross-border competence. One lax moment could be the difference between spending twelve hours horizontal or vertical or being left behind on the platform.

Of the sleeping berths (i.e. those that allow you to get horizontal), there are two types: the sleeper and the couchette. The sleeper is the top tier and the most luxurious, although luxurious is relative in this context. Typically, a sleeper cabin has three fixed beds stacked above each other on one side like a triple bunk bed, some modest storage and a sink that reminds me of those you see in prison cells on TV.

In contrast, the couchette cabin is more basic but not without its charms. The beds fold out so that during the day you can pack them away and revert to a normal train cabin with upright seats on either side. The beds are more basic than those provided in a sleeper, with the duvets and pillows sometimes just consisting of a sheet and a small cushion. They're often more cramped, as they tend to be four- or six-person compartments, but the ability to fold the beds away during the day can be beneficial on a long journey.

Alas, there is a change of plan. Our train has been replaced. Our sleeper cabin is now a six-person couchette, but we still have it to ourselves. The bed frames are a garish yellow and the upholstery on the seats is a deep blue, in a pattern that looks like a half-finished game of Tetris. The retro upholstery reminds me of the type you find on the London Tube. All in all, it is debatable as to whether it was designed with sleep in mind.

Budapest, Hungary – 29 October 2024

We arrive in Budapest at 11 a.m. to sunny blue skies and with eight hours until our next sleeper train departs. Built in the nineteenth century, the Budapest Keleti is probably the grandest station of the whole trip. It has a cavernous central hall adorned with the sort of frescoes that belong in a museum. The station's façade is even more imposing and looks like the gate to a medieval city.

After caffeinating (a constant throughout our trip), we stash our luggage and set off for the Gellért Baths, one of the many spas across the city, and to Jubileumi Park, which provides elevated views of the city skyline and the river Danube, which cuts the city in half. Following an excellent, if a little rushed, dinner at a small bistro in the back streets of the old town, we return to Keleti station for our next train.

In eight short hours, we've managed to visit a spa, see one of the best views of the city and enjoy a traditional meal. That, we tell ourselves, is what a journey like this is all about.

Striding confidently up to the platform for our 19.10 departure, we are met by a brusque and imposing ticket inspector from the Romanian national rail operator, which jointly runs this service with the Hungarian rail operator. My confidence fades when he doesn't recognise the format of our tickets. He reluctantly lets us on, and the train starts moving, but I can't relax. The ticket drama would either be resolved, or we could get booted off the train at the next station.

We're in a sleeper cabin with three beds on top of each other and a small metal sink. I step out into the old mahogany panelling of the corridor that runs alongside the sleeper cabins and am met by the ticket inspector. Things are not resolved. We survive thirty minutes of uncertainty before a ticket inspector from the Hungarian operator also arrives. Following a heated exchange in both Romanian and Hungarian, an agreement is reached that our tickets are valid. The first inspector reluctantly accepts this outcome, but we have several awkward encounters in that same corridor throughout the night. Finally, as I emerge from brushing my teeth in the bathroom at 7 a.m., his cold stare is replaced by a nod and a grunt.

Standing in the corridor, still quite early, we stop at a tiny station in Codlea, Romania. There's a short guard in a navy uniform and red hat standing on the platform; he marches comically backwards and blows an airhorn. It looks like a skit from a 1980s sketch show about the Soviet Union. I think it was real.

Brasov, Romania – 30 October 2024

It's 8.50 a.m. and we've made it to Transylvania. We don't need to stop here, but it is almost Halloween and this is the home of Dracula. Brasov is exactly the sort of place that we might not have found otherwise.

The train station makes a stark contrast to the one we departed from in Budapest. This one is a bleak grey and nothing about it belongs in a museum.

The medieval old town is a different story altogether. Centred on the Gothic-style Biserica Neagră, or Black Church, and surrounded by the southern Carpathian Mountains, it is a delight. We wander up to the Weavers' Bastion on the edge of the town, a fortification that once protected the city and now serves as a museum. It's early and is still closed, so we follow one of the many forest trails up into the Transylvania Alps, the autumnal leaves scattering the path in front of us.

The two-hour train from Brasov to Bucharest is even more impressive than Brasov itself, with most of the journey continuing through a forested valley of the Carpathian Mountains. The dark green spruces are overwhelmed by the burnt orange and dull yellow of the autumnal leaves.

Bucharest, Romania – 30 October 2024

Arriving in Bucharest, we head straight out to a grand beer hall in the centre of the city and find plenty of amusement, especially as the live band revs up, a dancefloor forms, and Jess is whisked away by one of the waiters in an upbeat waltz.

I wake up the next morning feeling nervous. This will be the most stressful day of the trip so far. We have a finely balanced schedule of train travel in front of us. Three separate trains criss-crossing their way from Bucharest, across the border into Bulgaria, and ending up in its ancient capital, Veliko Tarnovo. Miss one connection and there is no alternative.

The day does not disappoint.

The first train, from Bucharest to Ruse in Bulgaria, has the most jeopardy. The train is due to arrive in Ruse at 13.39 and our next train, to Gorna Oryahovitsa, is scheduled to depart at 14.15. The train waiting in the station is modern but covered in lime-green graffiti, making it almost impossible to discern the windows from the doors.

The journey takes just under three hours and I spend most of it glued to my phone, refreshing the online timetables to check timings. We were already delayed as we left Bucharest. My blood pressure begins to rise.

The real sticking point will be the border crossing and passport checks either side. As we approach the border, we are twenty minutes delayed. My blood pressure increases further.

The train rolls to a halt just ahead of the border. Two Romanian border guards amble towards the train, pausing briefly to finish their cigarettes. They make their way through the carriage collecting our passports, pausing occasionally for some unhelpful small talk, before disappearing to run their checks.

Precious minutes pass before they return, seemingly oblivious to the stakes for those on board. If they notice the pained expression on my face imploring them for more urgency, they don't acknowledge it, which is probably for the best.

Passports returned, we start to move again. As we snake across the Bridge of Friendship that crosses the Danube River and connects the Romanian bank to the Bulgarian one, I feel sick. It is unclear where or when the Bulgarian passport check will take place, but I know it is coming. Thirty minutes delayed. My blood pressure reaches unsafe levels.

We arrive on the platform at Gorna Oryahovitsa. There it is on the other side of the platform, the train we need to catch, once again covered in graffiti. But there *they* are as well, the Bulgarian border guards waiting for us outside the train doors.

They take our passports and disappear off, leaving a frenetic crowd in their wake. Where are they going? How long will they be? Will the train leave? The passengers can be split in two, the panicked tourists and the relaxed locals, who, perhaps unsurprisingly, all have the air of people who have seen this show before and know the ending.

The train should have left but it's still there. We may be in luck. One of the border guards returns and hands back passports to Jacques and me, but Jess's passport is still missing. More panic. Jacques and I run to the train to try and hold the door, still only 38 per cent certain that it definitely is the right train. Jess is remonstrating unsuccessfully with the border guards.

Finally, she gets her passport and races to the train, we get on and find our seats. For a few minutes, we feel like we've emerged victorious against the odds, then we discover that this is a daily occurrence, and the train never leaves until the first one has arrived and the passport checks are completed. The little things they don't tell you on the booking sites...

Veliko Tarnovo, Bulgaria – 31 October 2024

Veliko Tarnovo is probably the biggest surprise of the trip. When I planned the route, it was somewhere we had to stop because the direct train from Bucharest to Istanbul doesn't run at this time of year.

It was the capital of the Second Bulgarian Empire, we're reliably informed by the internet, and it turns out its history is almost as deep as the gorge that cuts through the old town. At the bottom of this gorge is the Yantra River and it separates the three hills that make up the old town.

One of these, known as Tsarevets, is home to the sprawling Tsarevets Fortress, which we can see lit up across the gorge from our accommodation on one of the other hills. This was our destination for the next morning, but not before a sunrise run to the Sveta Gora Park that looks back over the old town.

The fortress is sprawling and would have taken more time than we had to explore in full, so we clamber along the walls for a bit and check out the Ascension Cathedral, a medieval church at its centre. From the outside it looks like many other historical churches I've politely looked round as a tourist, but the inside is unlike any religious building I've ever seen.

The walls and ceiling are adorned with impressive but slightly unsettling modern gothic frescoes depicting groups of monks that look like the undead and everybody has overly angular facial features. Imagine the frescoes in the Sistine Chapel redrawn by a modern-day graffiti artist and you'd be close to what the Ascension Cathedral looks like.

Dimitrovgrad, Bulgaria – 1 November 2024

From Veliko Tarnovo, we are catching a 6 p.m. train to Dimitrovgrad near the Turkish border in southern Bulgaria. This is where we'll catch our third sleeper train to Istanbul.

It's pitch black when we arrive at the station in Dimitrovgrad, which is almost deserted save for a small gaggle of mostly British travellers attempting the same transition as us. There's not full panic among the group, but there is a great deal of anxious uncertainty, as it's unclear when or if the train is arriving. Our tickets say one time – 23:33, the board in the station says something different, and the internet offers up a third option. Every time the shadowy station is lit up by the lights of an arriving train, the group rushes out into the cold of the platform in anticipation. This happens several times before the train eventually arrives at a different time to all three of the above.

During the planning for this trip, I'd read a lot about this train and I knew it would be taxing on our moods. It is already past midnight, and we have two border checks to go through before we can sleep. I'd also read that on the Turkish side, every passenger needs to get off with their luggage to go through passport control and customs, so we shouldn't unpack or get into our sleeper berths. Within five minutes, despite our warnings, Jess proceeds to unpack and then clambers under her duvet.

The train stops at the Bulgarian border point at Svilengrad. This one is easy enough, with two border guards getting on the train to take our passports for checks. That makes sense – we're leaving Bulgaria after all and border checks are always easier in the country you're leaving.

The train continues slowly to Kapıkule station just across the border in Turkey. By now it's close to 2 a.m. Sure enough, we're instructed to get off with our luggage and join the queue for passport control, which is located in an aggressively lit room. There is some dim chatter, but most people are shuffling forward in a bleary-eyed silence.

Back outside and to another room to put our luggage through an airport-style scanner before we're held on the platform while the train is searched. We notice a dilapidated sign for duty-free, which amuses Jacques and me as we contemplate getting something for the rest of the journey. Jess says no.

We get back on the train, half-heartedly unpack the essentials and collapse in our sleeper berths. As ever, Jacques and Jess are on the top two and I'm on one of the bottom berths.

Istanbul, Turkey – 2 November 2024

The Orient Express was the original long-distance sleeper train, and its name still captures the excitement of a trip like this. Sirkeci station in central Istanbul was the eastern destination of the original route until 1977.

Pulling into Sirkeci station on our rather less glamorous train, slightly dazed by the nighttime antics at the border, we are unprepared for the frenetic nature of Istanbul.

This is the sixth day of travel and, with two nights before we move on, Istanbul will serve as a break on our trip to rest up, get some work done and explore.

We wander around the bazars and central squares, go to the Blue Mosque and drink Turkish coffee. Everything feels confusing but everyone is friendly. We find ourselves shuffling along a busy street before heading into an even busier market in search of lunch. We find somewhere and sit down for a very satisfying meal, even if it's never quite certain what we're ordering, when it will arrive or how much it will cost.

We are staying in Kadıköy on the eastern side of Istanbul. I'd chosen it as our base because it was close to the train station we'd depart from but our journey to Kadıköy itself turns out to be one of the best things we do in Istanbul. We take one of the many commuter ferries across the Bosphorus at dusk, where East meets West, as everyone likes to say. Looking out from the back of the boat, a bright red Turkish flag fluttering in the wind, you get a superb view of the Istanbul skyline. It's quite distinct from the European cities we've been travelling through. Minarets from the many mosques shoot up across the city's skyline, guarding the domes that sit at the centre of each mosque.

Jess embraces the journey and orders some tea, which arrives on a silver platter carried by the only man who doesn't wobble as he navigates the boat deck. The tea is not the best we have in Istanbul, but it is the one that comes with the best view.

Ankara, Turkey – 4 November 2024

Two nights in Istanbul and it is time to head to Ankara. It's a four-hour journey on one of Turkey's flagship YHT high-speed trains. The Istanbul–Ankara route opened in 2009 and is one of only a handful of high-speed routes in the country. It is noticeably more modern than the other trains we take in Turkey, and we get a lot of work done in our spacious air-conditioned carriage.

Arriving in the capital, it feels smaller and more business-like, but then most places probably would compared to the size and chaos of Istanbul. We head to a place for dinner down the road from our hotel, where the waiters put on a show. Each dish comes with a serving protocol that ranges from steaming saucepans to intimidating carving displays. The only thing more intimidating than the carving is the spice, which I am unprepared for, but the meal is so good we return for lunch the next day.

I go for a morning run up to the Ankara Citadel, the old town and original fortifications of the city built around the seventh-century Ankara Castle. It feels like somewhere that would normally be bustling with tourists, but it's still dark and the city is yet to wake up, so the only company I have is a collection of stray dogs.

Post-run, we don't have loads of time before heading to the station, so we pick our exploration point carefully. We go for Anıtkabir, a huge mausoleum to the first president of the Turkish Republic, Mustafa Atatürk. It sits atop a hill in the centre of the city and is surrounded by well-manicured gardens. The building itself is striking: a large rectangular building with pillars across the façade and Turkish flags on either side fluttering in the wind against the clear blue sky. It reminds me of a fully formed, more angular version of the Parthenon in Athens. As we get closer, the resemblance is more Lincoln Memorial, with Atatürk's speech about Turkish independence in place of Lincoln's Gettysburg Address.

From the West to the East, Turkey – 5 November 2024

Up next is the most anticipated part of the trip – the Doğu Express, or Doğu Ekspresi, which means Eastern Express in English. We'd taken to calling it the Big Bad Doğu, for no other reason than this seemed to capture our excitement and incite it further.

The Doğu is a twenty-six-hour sleeper train connecting Ankara in central Turkey to Kars in the east. We'd be getting off slightly earlier, in the mountain town of Erzurum. This meant that our journey would be just twenty-two and a half hours, which I was feeling pretty smug about until I learned that it meant we would miss out on the kebab refuel stop – an ingenious innovation where the train staff take orders from passengers and have fresh kebabs waiting to come on board at the next station. If I were Railways Tzar for a day, I'd make this mandatory on all trains.

The Doğu Express was built in the 1930s as a commuter train for locals to get across the vastness of the Turkish countryside, but in recent years it has become a tourist attraction, its popularity turbocharged by social media influencers sharing videos of the picturesque snow-capped mountains of the Anatolian Plateau. It's now so popular that the Turkish State Railway company launched a tourist version that runs during the winter months. It takes slightly longer, at thirty-six hours, and has many more stops along the way. However, this option was unavailable at the time of our trip, so only the original train was running, making our experience seem slightly more authentic. This didn't make it any easier to get tickets, because they sell out almost as soon as they're released. The fact that we ended up with sleeper berths rather than upright airplane seats on our twenty-two-and-a-half-hour train journey means that I will forever be in debt to a Turkish travel agent named Olga.

The boarding process at Ankara's Garı station is slightly underwhelming. We take pictures on the platform and Jacques makes a dash for a couple of celebratory beers.

The experience soon starts to match our expectations as we whoosh through the Ankara suburbs with the sun setting. We sit in our cabin and toast the Big Bad Doğu with our first beer. A few minutes pass until we are politely informed that we shouldn't bring our own alcohol on the train, so we apologise and quickly down our second beer.

Again, we are in a four-person couchette. The upholstery is a dark green in an indeterminate pattern, a much more sedate choice than our earlier sleepers.

The experience of waking up on the Doğu the next morning is easily the best of the trip. The sun slowly rises, revealing seemingly endless golden fields and the rocky outcrops of a mountain range we are soon in the middle of. The centre of each train window is stencilled with the white star and crescent moon of the Turkish flag, adding to the magic of the views outside.

We make coffee in our cabin to fill our trusty keep cups before decamping to the ramshackle dining car, a collection of burgundy leather seats around school canteen tables. Coffee is available in the dining car, but we have with us our own mini-travel kettle, some coffee we'd picked up in Veliko Tarnovo and a pretentious coffee drip-filter thingy. It is fun to use and makes for excellent social media content. We are, after all, on the premier influencer train journey this side of the Bosphorus (and maybe on the other side, as well).

The train doesn't seem that busy. There are quite a few locals making their regular west to east journeys and a few couples that appear to be holiday with their big hiking backpacks. We meet a German journalist also attempting to make her way to COP by land and a young Englishman travelling by himself as a break from his corporate job back in London.

We drink coffee, do some work, chat, drink more coffee, play cards and stare out the window. The landscape is beige in tone but mesmerising. We skip between mountain passes and emerge onto the vast open Anatolian plains. At one point we find our train running alongside a turquoise river as we both cut through a deep gorge. Out in the open, the sky is so clear that every time we pass a body of water, we see the surrounding mountains perfectly reflected back at us.

We approach our stop and pack up our stuff but not before one last look at the scenery out the window. It has been gloriously sunny all day and I've never had a train journey quite like it.

As we alight from the train, I keep my head down so as not to see the fresh kebabs being delivered to our acquaintances still on board. I know what I'll be ordering for my dinner.

Erzurum, Turkey – 6 November 2024

Erzurum is almost 2,000 metres above sea level and surrounded by mountains in eastern Anatolia. We arrive at dusk and are met by a wall of cold mountain air. Jacques and I have been preparing for this moment for some time, and have transitioned to trousers, thick coats, gloves and woolly hats. Jess hasn't taken our warnings seriously and is taken aback by the biting air.

We satisfy our food cravings and wander around the town. It's a jarring combination of semi-modern, nondescript buildings and busy shopping centres with a handful of beautiful monuments and madrasas, dating back to the twelfth and thirteenth centuries. The most impressive of these is the Çifte Minareli Medrese, or Double Minaret Madrasah. By now it's dark, and the two minarets are lit up against the crisp mountain night with the faint outline of the mountain peaks behind them. Jess wants us to pose for photos in front of it, but we're too cold for a photoshoot.

We head back to our hotel but not before procuring some simit, a soft Turkish bagel covered in sesame seeds that we'd become addicted to in Istanbul, for our journey the next day.

Sarp/Sarpi, Turkey/Georgia – 7 November 2024

Another day, another border. This will be our first crossing on foot. Our destination is the primary Turkey–Georgia border crossing on the coast of the Black Sea to the northeast of Turkey. The border is busy, with a long line of lorries and cars snaking back along the single coastal road that leads to the crossing point. Those on foot are mostly families, pushing large trollies of luggage and furniture.

We were unsure how easy this crossing would be, but it turns out to be quick. We even have time for coffee in the no-man's land between each of the two border controls, and for Jacques and me to finally look in a Turkish duty-free.

Batumi, Georgia – 7 November 2024

On the edge of the Black Sea, about twenty kilometres from the Turkish border, Batumi is quite unlike any major city I've been to. It's the second largest in Georgia and it seems to be in the midst of an identity crisis.

Dubai-style skyscrapers crop up across the skyline, while Las Vegas-style casinos pepper the central squares, and the beachfront feels like an English seaside town, with a pebble beach and aged fairground rides. All of this surrounds a well-maintained old town that has a slightly Mediterranean feel, with colourful façades, quirky balconies and even the occasional palm tree.

On the opposite side to the Black Sea, mountains are carpeted in thick green forest.

There's clearly a lot of money in the city, and as I go for my morning run most of it seems to be falling out of the casinos that surround the Central Europe square.

Before our evening train, we decide to explore the botanical garden we spot on the map, just to the north of the city along the coast. The garden is steep and the weather is warm, but we are rewarded with great views back to the city and out across the Black Sea.

The 5 p.m. train from Batumi to Georgia is the final one we have pre-booked and the last part of our journey that is certain. The train snakes slowly up the coast along the edge of the Black Sea and then straight through the middle of the botanical gardens before swinging right and inland towards Tbilisi.

The train is modern, comfortable and busy. The plush leather seats are filled with tourists heading home. Some of them seem familiar from outside the casino that morning.

Tbilisi, Georgia – 8 and 9 November 2024

The first thing that strikes me about Tbilisi is the graffiti. Not the volume, every city has its graffiti, but the subject matter. I've never seen graffiti extolling the virtues of the North Atlantic Treaty Organization (NATO) before. Georgia has just been through a bruising election, in which a populist party was returned to power on a pro-Russia and anti-EU platform. The streets have witnessed lively protests in the previous days, but it seems fairly calm when we arrive.

Moving out and uphill from the Kura River that runs through the centre, Tbilisi is a hilly maze of a city. Just behind our hotel to the south-west of the city lies Mount Mtatsminda. We clamber up it the next morning, pass the Mtatsminda

Pantheon, a necropolis for worthies past, to look back down over the city as the sun rises. The sunrise on the Doğu Express is hard to beat but this one comes close.

Jacques and Jess, intrepid wild swimmers in all locations and conditions, find a secluded lake up on the mountain and go for a swim. I do not partake and decide to join them back down below for the coffee-and-cheesecake scene, which is excellent.

As fun as Tbilisi is, the issue of the closed Azerbaijan border hangs over us, so it is with mixed emotions that we head out Saturday night for a drink.

Every time we've been asked about the closed border, we've repeated the same mantra. Like a politician who's not quite ready to admit the truth:

'We have submitted a formal application to the Cabinet of Ministers of the Government of the Republic of Azerbaijan for special permission to enter via the land border. We are in contact with the Embassy and hope to hear soon.'

The more we've said it, the more absurd it's sounded.

But now, as we sit in our Georgian purgatory, the sun starting to set as the buzz in the bar starts to rise, the email arrives.

Just three sentences. The first two:

We are pleased to inform you that permission has been granted for Jack Curtis, Jacques Sheehan and Jessica Rogers to enter and exit Azerbaijan through land borders. The reference number for this permission is 1-5/1-17128/2024.

We've done it. We have permission. We can complete our trip by train and without flying. But wait, where is the border crossing? How do we do it? What do we need to bring?

The third and final sentence of the email reads:

Please note that until November 14, 2024, there are non-working days at the Consular section of the Embassy. Consequently, inquiries will not be responded to during this period.

We had assumed that the border crossing was at the point where the dormant Tbilisi–Baku sleeper train crossed, but some last-minute googling reveals that we should (probably) head to a different point, the 'Red Bridge' border crossing.

There is nobody for us to check with, no way of knowing for certain, so, once again, we decide to roll the dice and hope our plan will work.

The Georgia–Azerbaijan border – 10 November 2024

We wake up with a frantic energy and soon we're pulling up to the border. The first thing we see are some men fighting. Not a good start. We get out with our bags and walk towards the Georgian border guards. They look at us quizzically. Nobody is meant to cross this way. We explain our case and hand over our paperwork, including the three-line email with our special permission.

One of the guards walks away with our paperwork and passports to start making some calls. His colleagues come over to chat. They seem just as bemused by our arrival as the first man. Eventually, he waves us through but the look on his face suggests he thinks we'll be back in a few minutes. We walk to the Georgian passport control. Again, always easier on the way out…

Now we're in no-man's land. We've left Georgia, exit stamps in our passports, and we're not quite sure what we're walking towards.

We get to the Azerbaijan border office and a soldier is waiting for us. He's a giant of a man in full army fatigues and a big moustache. He's expecting us, but it's much less clear if he was expecting our arrival before he was called by the Georgian border guard ten minutes earlier. It is pretty remarkable that we were granted permission, and we don't know of anybody else that has been so favoured, so he may well have been notified in advance. His boots were certainly well polished for the occasion.

We get our passports stamped by a border guard, our visas checked and our bags put through security scanners. We seem to be through. The soldier is back and asks for our passports again. We oblige. He enters a room off to the side and leaves the door ajar. He proceeds to take photos of our passports on his phone and makes some calls. We'd been through the official passport and visa checks. This is unofficial. I don't know what those photos are for or where they will be sent, but in that moment I resolve to not so much as jaywalk while I am in Azerbaijan.

My nerves have barely settled when we spot a car waiting for us on the other side, spluttering black smoke, a crack thundering across its windscreen and a driver hellbent on taking us eight hours all the way to Baku for a small fortune instead of the train station we need, which is in Aghstafa, about fifty kilometres away and our closest. Jacques takes on the negotiations; he's good in these situations, and keeps checking the map app on his phone to make sure we are heading to the station.

Aghstafa, Azerbaijan – 10 November 2024

It's 3 p.m. by the time we make it to Aghstafa. There is one train left that day to Baku, the capital and our ultimate destination. Our final train of the trip. We are jubilant.

We stand on the platform in a state of disbelief. We are an anomaly and there is some interest in our presence. I go to get tickets for the train, while Jacques and Jess make conversation with a friendly policeman. He doesn't speak a word of English, and they don't speak a word of Azerbaijani, but they must have made quite an impression because he returns ten minutes later to present Jess with flowers.

At 5 p.m. our train arrives and we hop on, exhausted and happy. It's quiet and the journey is a long five hours. As we sit around our table, a teenage boy sitting across from us strikes up a conversation in English. He wants to know what we're doing on a train from Aghstafa to Baku at 5 p.m. on a Sunday. It's a fair question. We tell him and he starts smiling. He's also heading to the COP conference, as a volunteer.

Baku, Azerbaijan – 10 November 2024

We arrive in Baku just after 10 p.m. on Sunday, 10 November. The COP29 climate conference starts the next day.

We've been travelling for two weeks and taken fifteen trains over 6,000 kilometres through ten countries. London to Baku by train. From western to eastern Europe, to where Europe meets Asia, across the length of Turkey and into the Caucasus.

Our journey to get to Baku was extreme, but it was meant to be. It's symbolic, to capture the imagination and show that sometimes the train is an option.

Global aviation emissions are growing, and while flying is still necessary in many situations, changing our behaviour can help limit its impact. The carbon emissions from our journey were twenty times lower than if we'd flown to Baku. I've done several of these trips and the emissions are always significantly lower, and the difference will only improve as more and more trains are electrified. Avoiding a flight if you can is one of the most impactful decisions you can make as an individual.

Deciding whether to take the train or the plane is about trade-offs. Our journey was better for the planet, but it did take longer. A lot longer. It did cost more, but only slightly more per person than direct flights, such are the benefits of booking in advance and using Interrail passes. All in all, on this occasion, it was a trade-off we could make.

Sometimes the train will be too expensive, or take too long, but there will be many other times when the trade-off will be easier. Taking the train from the UK to the south of France instead of flying, which is a journey that can be done in a day and for little expense, is twenty-nine times lower in terms of emissions. The train is often a more pleasant and easy journey than the chaos and restrictions of flying, and is far more likely to offer up unexpected delights and unforgettable memories. It might not even feel like a trade-off at all.

Delivering humanitarian aid from the Ukrainian Railways
Food Train.

IRON BRAVERY
FELICITY SPECTOR

How Ukrainian Railways keep running

At six in the morning, three-and-a-half years into Russia's full-scale war, the railway station in the eastern Ukrainian city of Sumy was not exactly bustling. But passengers had begun arriving on the narrow platform for the train to Kyiv, while a small shop was doing a roaring trade in small paper cups of strong coffee for sixteen hryvnias, around thirty pence. Russian forces were less than twenty miles away: at night the sky lit up with incoming drones, and the city had been increasingly hit by cluster munitions and guided air bombs. But the trains still serviced Sumy station every day, arriving and departing exactly on schedule.

The coupe compartment with its four bunks was neat and clean, with packages of freshly ironed sheets and pillowcases for each passenger. The conductor came round, smiling, offering a choice of tea. The actual route to Kyiv was more of a surprise, passing within a few miles of the Russian border through villages under active evacuation. We chugged slowly through Bilopillia, where a week earlier nine people had been killed when a Russian drone hit their evacuation bus. Every window in the station building was broken and boarded up, but the platform was immaculate. At the next village, two men got on, loaded with suitcases and bags. They piled everything into the top bunks and took up residence on the lower ones, scanning the air-raid apps on their phones, which buzzed constantly with alerts and which translates from Ukrainian as 'anxiety'. 'What's it saying, is there a ballistic threat?' 'Yes, missiles from Kursk region. But not in this direction.' The train sped up slightly towards Konotop and onwards to Kyiv.

The following night, Ukrainian Railways reported a drone strike which had damaged part of the track in the Sumy region, but it continued proudly: 'traffic in this section was not subject to delay'. Driving so close to an active combat zone, though battered villages where residents are advised to flee for their own safety and under treacherous skies, has become part of life for the courageous staff of Ukrzaliznytsia. Too many have lost their lives to Russia's war. Yet they still manage to keep the trains running on time. And this is what they call Iron Bravery, the extraordinary story of what a company can achieve during war.

When the big invasion broke out, the firm's senior management team were meeting in Kyiv and quickly dismissed any suggestion that they should leave for safer territory in the west of Ukraine. They would not see their homes again for months. In an interview at the time, the then CEO, Oleksandr Kamyshin, said they realised that Ukrzaliznytsia was one of the most vital parts of the country's infrastructure. 'People should rely at least on something,' he said. 'They rely on the army, they rely on the President and they rely on the railway.

We should not cancel a single train.'[1] Bear in mind that at this time, Russian missiles were exploding over Ukrainian cities, including Kyiv: warplanes were in the sky, while vast columns of enemy troops were pouring over the border backed by artillery and tanks. As millions of Ukrainian families rushed to flee, the roads were perilously clogged with traffic. The fastest and least dangerous way to evacuate such an overwhelming volume of people was by train.

Oleksandr Pertsovskyi was barely into his mid-thirties, but as head of passenger services it was his job to organise this unprecedented evacuation plan, not only keeping trains running but getting vast numbers of women and children safely out of the most perilous zones. A friend of mine remembers what it was like in those terrifying days: she had taken her parents to Kharkiv station, which was so crowded there was barely room to stand. She found a corner where her mother could sit on the floor, while her father remained standing all night, holding tightly on to their cat. Carriages meant to hold fifty-four passengers were instead crammed with up to a hundred. According to *Forbes*, in the first month of the war, the company managed to transport three million people westwards, 400,000 of them abroad.[2]

The most comprehensive book about this extraordinary period is *The Train Arrives on Time* by the Ukrainian journalist Maria Paplauskaite. She describes in intimate detail how the five men at the top of the company first sent their families out of the capital and then lived constantly on the move, on board

1   https://www.businessinsider.com/how-ukraine-lifeline-railway-runs-even-with-russia-bombing-it-2023-2
2   https://forbes.ua/lifestyle/urok-iz-krizovogo-menedzhmentu-yak-kamishin-ta-shche-5-lideriv-ukrzaliznitsi-zberegli-kontrol-u-pershi-dni-vtorgnennya-urivok-iz-knigi-potyag-pribuvae-za-rozkladom-20042024-20663

a single-car diesel command train called Motrysa. They travelled to evacuation hubs and strategic locations, coordinating everything through a Soviet-era closed telephone network which linked them to the system's 1,450 stations. It was assumed that the company executives were on a Russian 'kill list' and so, as they said in interviews at the time: 'Move fast, so you don't get caught. And don't spend more than a few minutes in each place.' They divided up tasks between themselves, covering everything from passenger safety and fuel supply to security and communications.[3]

This most extreme form of crisis management was all about the passengers and keeping the entire country moving, hour by critical hour. For Oleksandr Kamyshin, there was an overwhelming imperative: reliability and resilience. He even coined a hashtag for it, which became popular on social media – he called it #KeepRunning. The highly distressing scenes from that time resemble footage from the Second World War. Railway platforms crammed with vast and desperate crowds. Children's faces pressed against windows, as they were ferried away from their fathers, who were unable to leave the country under martial law and had to stay behind. Possessions were reduced to a single backpack, a carrier bag filled with the few things most important in life. In these frantic, perilous weeks, hundreds of thousands of people a day were transported by train, without knowing when they might see their homes and families again. At the heart of it all were Ukrzaliznytsia's 200,000 employees – almost half of them women – and such was their fortitude that they became known as Iron People.

---

3   https://forbes.ua/ru/company/neefektivna-ukrzaliznitsya-peretvorilasya
-na-mashinu-iz-poryatunku-milyoniv-ukraintsiv-yak-kompaniyu-perestavili
-na-voenni-reyki-04042022-5255

This mass evacuation of millions of people and transport-ing critical humanitarian aid under fire would be more than enough for any company to cope with – let alone one whose senior executives were operating from on board a constantly moving train. But Kamyshin and his team often took a personal role in helping passengers out – as I discovered for myself in December 2022. There had been a particularly large-scale Russian missile attack on the capital, Kyiv, which had taken out much of the city's electricity supply. The metro was out of action, traffic lights did not work, and internet access was patchy at best. I was booked on a train to Poland that evening, and when I managed to log on to Twitter (as it was then) I was surprised to see a direct message from Kamyshin himself. 'Are you booked on an inter-city train tonight?' he asked. 'It won't be running because of the electricity problems. But we can help you change tickets onto another service.' He sent me the number of someone in the team who would assist, and a series of instructions arrived on what to do. Kyiv central train station was plunged into darkness, and none of the escalators worked, but I followed the instructions and found a small office lit by a battery-operated lamp, where two women managed to issue a ticket on an alternative overnight train. They insisted on refunding my original ticket, leading me down a maze of shadowy corridors to another room filled with people crowded around a single laptop plugged into a power bank. Kamyshin sent me frequent messages: asking if I had the new ticket, if I was on the train, if I had any comments or feedback about the journey. It was the kind of hands-on, personal involvement that I was to witness frequently among the top management at the firm.

Another sign of resilience was the speed at which services

were restored to de-occupied areas. 'First our tanks go in, then the trains,' Kamyshin said: and you could see that happening in places like Balakliya in Kharkiv region, where trains started running the day after the place was liberated. In Kyiv train station, there was a mocked-up departure board with the names of all the cities under occupation where one day Ukrainian Railways would travel again. The first of these services to be restored was the one to Kherson, shortly after Russian forces retreated to the opposite bank of the Dnipro river. The city was not to enjoy a peaceful life, however: from the very first days of liberation it was mercilessly pounded by Russian artillery and drones, including the railway station and trains. Maria Paplauskaite came under one such attack when she was in Kherson to interview railway staff for her book. She describes running for cover in the station shelter together with passengers who had just been trying to board a train – a police officer was killed as he rushed to help, while five others were injured in the shelling.[4]

By the end of 2024, some 729 railway employees had been killed, with more than 2,300 injured – while many hundreds more had lost their homes. The company set up its own organisation called Iron Families, to help survivors and bereaved relatives with everything from psychological to financial support.

After the initial crisis of mass evacuations receded, Ukrainian Railways turned to other innovative ways to support a country at war. Their first project was to build medical evacuation trains, to take injured or seriously

---

4   https://suspilne.media/culture/774969-potag-pribuvae-za-rozkladom -vse-so-varto-znati-pro-novu-knizku-vid-maricki-paplauskajte-ta-vidavnictva -laboratoria/

ill people from frontline areas where there was little or no access to acute hospital care. The idea of using trains was not entirely new – the Americans had transported injured soldiers by train during the Korean War. It made especial sense in Ukraine, with its vast distances and bad roads, especially in the eastern Donetsk region, where people would tell you that the shocking state of the roads was an endemic problem: 'This is corruption, it is nothing to do with war.' The railway company partnered with the charity Medecins Sans Frontiers (MSF) to build fully equipped clinics inside train carriages, able to triage and provide basic medical care to patients, on the move, until they could reach hospital. The first carriages were brought into service astonishingly fast: at the time, Kamyshin recalled that it would normally take a year to create such a project, but they managed it in just four days. Two years after the full-scale invasion, seventy-eight carriages had been kitted out for medical evacuation.

The next challenge was building a fully equipped ICU for the most seriously ill patients: the carriages needed their own generators and water purification systems, while there were major logistical issues around oxygen tanks, which could be a high explosion risk if the train were to come under Russian shelling. Even the speed of the train had to be modified – travelling at around half the usual speed to make it more stable for medical staff carrying out even the most routine proced-ures, like blood transfusions and cardiac monitoring. Most of the medics were locally hired, from those with experience in emergency medicine to technical logistics, while Ukrainian Railways employees kept the trains going. It was vital for safety reasons to keep the route and location of the trains secret, while patients had to be taken off and loaded into ambulances with incredible speed.[5] The MSF charity spoke to some of the staff involved in this incredibly stressful work, including a woman called Nataliia at Kherson station, who described how she tried to keep distressed civilians calm, even while explosions were going off nearby. 'After each evacuation,' she told the charity, 'there are patients whose eyes I remember for a long time. I ponder their stories, their paths and destinies.'[6]

The next project came about as a result of a partnership with the American philanthropist Howard Buffett, son of Warren Buffett, one of the world's richest men. Howard Buffett had become heavily involved in Ukraine, travelling frequently to frontline areas and donating hundreds of millions of dollars of his own money to food security and reconstruction efforts. Ukrainian Railways came up with the idea of creating the world's first Food Train, creating an entirely self-contained travelling kitchen inside specially retrofitted carriages. With funding from the Howard Buffett Foundation, Ukrainian engineers began work in a railway depot, normally used for repairs and modernisation, converting three passenger carriages and three cargo cars into hot and cold kitchens, storage and refrigeration, and accommodation for the staff who would work on board, complete with shower and washing machine. 'This was just a depot, not a big factory,' they told me. 'It had limited space; there were many technical connections to create: it wasn't done in a day.' Like the medevac trains, it was also equipped with a four hundred kilowatt generator and a dedicated space for 27,000 litres of water, complete with a filtration and pumping system.

5   https://www.doctorswithoutborders.org/latest/36-hours-aboard-msf-medical-train-ukraine
6   https://pmc.ncbi.nlm.nih.gov/articles/PMC10290241

The Food Train began working in 2023, designed to operate independently for up to a week: it even had its own refuelling point to keep the generator running. It would be able to bring thousands of hot meals every day to areas worst affected by the catastrophic energy blackouts caused by Russia's relentless attacks on civilian infrastructure.

Railway stations, especially in places near the front line, had already become home to so-called 'indestructibility points', where people could shelter in a warm place, charge their devices, and get cups of hot tea or coffee. 'We wanted to provide people with more than just hot tea and a mobile connection,' press officer Oleksandr Shevchenko told me. Pre-packed ingredients arrived, courtesy of another partner – the Public Health Foundation. Twelve kitchen staff worked alongside eight employees from the railway, getting up in the early hours of the morning to prepare and portion out up to twelve thousand meals a day, stopping for a quick break at lunchtime and then carrying on until around six in the evening.

The kitchen staff were supplied by a company called Set Catering, run by the dynamic Oleksii Shpionov, a long-time partner of World Central Kitchen. His wife, Maria, also worked on board, as the team manager. Every afternoon, the chef in charge would draw up a spreadsheet of the organisations that had ordered meals for the following day, so he could calculate the menu and portions. It would appear in the team WhatsApp group just after midnight, a revolving menu focusing on around ten basic ingredients. There would sometimes be a soup, always a salad, some kind of meat, a grain like buckwheat or rice. Everything was boxed up and labelled for the charities and hospitals that had put in orders, so that their cars could pick it up at a pre-arranged time, driving right up alongside the train.

There were the same necessary safety precautions as used for the military evacuation service: 'The point of this train is to come as close as possible to the front cities. So we can't disclose the exact location until after the volunteers and social organisations have come to collect the food.' With typical dark humour, the team revealed that it was commonly known as the Doomsday Train. 'Because whenever something bad happens in Ukraine, it's ready to go. We would be happy if it didn't have to run at all; that would be a perfect situation if it wasn't needed.' The first time I visited the train in Kharkiv, it was busy making meals for the city's main psychiatric hospital, which had been hit by two S-300 missiles the day before, damaging the facility's kitchen area. The hospital's exhausted staff were responsible for seven hundred patients, who had nowhere else to go. Thanks to the Food Train, said one of the managers, at least they didn't have to worry about keeping the patients fed with hot and nourishing food.

Artur Botchenko, the young project manager of the Food Train, was also involved in the work of the evacuation trains, which carried thousands of families away from the hottest parts of the country towards relative safety further west. With their trademark determination, Ukrainian Railways refused to abandon these frontline cities until there was absolutely no alternative. The last train evacuating passengers from Pokrovsk, just a handful of miles from the front line in Donetsk region, left at the beginning of September 2024. By then, the city was coming under relentless shelling, and the company conceded that even 'gathering people at the station has already become very dangerous'. Figures collected by the World Bank in January 2025 showed that at least 126 stations and more than three hundred miles of track had been damaged by Russian attacks. In monetary terms, the amount of damage was more than $4 billion.

Sometimes those attacks were so horrific they made global headlines – like the bombing of Kramatorsk station in April 2022, which left at least fifty-eight civilians dead, nine of them children – and a hundred others injured. Despite Russia's initial denial, debris of the Tochka-U rocket could be clearly seen at the site, with Russian lettering scrawled on the side. President Zelenskyy called it 'an evil which has no limits'. However shocking and relentless the attacks, the railway company found ways of repairing the damaged infrastructure at astonishing speed. A typical message from communications chief Oleksander Shevchenko would read like this one, from August 2024: 'On behalf of all railway workers, we apologise for the inconvenience, because the night time shelling affected the plans of thousands of Ukrainians. We understand this and will do our best to get you to the desired destination as quickly as possible.' On that particular night, after railway infrastructure in the Poltava region came under fire, damaged sections of track were repaired within hours. Several services were delayed for between one and three hours, and others had to take a significant detour – but passengers whose journeys were disrupted were assured that they would be given free lunches prepared by the World Central Kitchen charity. On other occasions, when overhead power lines were damaged, long-distance services managed to keep running with back-up diesel locomotives. There always had to be a Plan B on Ukrainian Railways.

In May 2025, a day after Kherson station had come under attack yet again, damaging the station building as well as train carriages, I boarded the night service to Kherson from Lviv. It was brightly painted with the words 'Victory Train' along with scenes of crowds waving Ukrainian flags. One carriage bore the words 'Russia is a terrorist state' in English and Ukrainian. The violent attack the day before had not intimidated the railway company in the slightest. Their response was worded in typical Keep Running style: 'after the shelling ended, the train moved on with a delay of about an hour. The movement of other trains continues on schedule.' Service on board carried on as usual, the bag of fresh linen for the bunk bed, the choice of herbal tea, the polite knock on the compartment door to rouse you from sleep half an hour before your destination. The train driver continued his route according to the timetable, the conductor was stern in her neat uniform. Inside the carriage, you could see a soldier helping an elderly lady up the steps with her bag, and a mother tucking her young son into a top bunk, wearing odd socks, one blue and one yellow, the colours of the Ukrainian flag.

And still, every life lost was a heartbreaking tragedy. On 7 June 2025, Kharkiv came under attack in broad daylight with drones, missiles and guided air bombs. One of the bombs, known as a KAB, smashed into a building belonging to the Children's Railway, an educational project to give young people an engaging way to learn about the train system. The Kharkiv location was not open to the public, because of the city's high risk, but a young woman called Hanna Demenkova, who was thirty years old and had been appointed to lead the project just two weeks earlier, was killed in the blast. Another employee was so badly injured he failed to recover. Four others were treated in hospital, two in critical condition. The pink building of the Children's Railway depot was in ruins, the ground covered in shrapnel and broken glass and the bodies of victims – already a crime scene.

Theirs are the untold stories of Ukrainian Railways during Russia's full-scale war. A company that still finds time to expand its network and to build partnerships abroad, which boasts a punctuality record as high as ninety-seven per cent. A company that joined forces with the Howard Buffett Foundation a second time, turning part of a train into a mobile art gallery which would spend four months touring the country with a photography exhibition, depicting heroic Ukrainians on the front line.

These are the stories of staff who daily turn up to work on time, even when their job involves evacuating passengers under shelling, caring for patients in a moving ICU, or delivering essential food aid to cities on the front line. The workers in the darkness of night who repair broken tracks and fix carriages and rolling stock, who turn out at funerals to mourn the colleagues they have lost. They are the railway company that Ukrainians can rely on to get them where they need to go. In Oleksandr Kamyshin's words: 'We will stand for as long as it takes. And we don't just stand. We approach victory.'

Passengers wait to trundle down the coastline on the Sabah State Railway.

THE NORTH BORNEO RAILWAY

OMAR MUSA

Travels to the interior

Passengers rise from their seats in the small cafe situated at the entrance to the Sabah State Railway station in Sembulan, picking up their fragrant nasi lemak sold in brown paper pyramids and their sweating 100 Plus cans, to crowd onto the platform. Uncles wearing baseball caps or songkok, chatty teenage boys making teenage-boy jokes, mums wearing colourful tudung and shouldering babies. The crowd jostles forward as one, sensing the arrival of the train. It's a skinny, low platform, but above, grandiose lavender clouds pattern a peachy tropical sky. Lathered in perspiration, I remain crouched on the platform. There's no way the train will be on time – everything's always running late in Sabah, where my father was born and where I moved two years ago, to live here for the first time in my life.

After-work traffic hoots and honks. Sembulan station is situated at a busy intersection of Kota Kinabalu, the biggest city in Sabah, Malaysian Borneo's easternmost state.

Surrounding the station are a football field, manicured palm trees, a scraggly car park, government offices and, opposite, the enormous Sabah State Mosque, constructed in the 1970s, with gold minarets and domes, the largest of which sports a mosaic of gold and gunship-grey tiles.

The train grumbles into the station, blue and white, diesel-powered. And amazingly – on time.

It's a bunfight to get on, but I battle for position, and in no time I've secured a seat next to a chronic man-spreader (takes one to know one, I guess). As far as I can tell, the train is full of commuters who work in the city but live in villages and towns dotted along the coastline. My sweat cools as the air-conditioning blasts. The interior of the carriage is retro – a faded cream with aqua and sea-green plastic seats. I exhale, thinking about the diagnosis I've just received at the hospital not far from here, then sink into the seat.

At 5.30 p.m., just as the schedule said it would, the train creaks out of the station. It's quiet now, the late-afternoon calm punctuated by the giggles of a circle of young women who sit cross-legged on the ground next to the door, and the uncle in front of me flicking through endless YouTube Shorts on his phone.

This is the only train line on the whole island of Borneo, the third largest island in the world, or fourth largest, if you include the island-continent of my birth – Australia. Situated on the equator, Borneo is the only island in the world shared by three nations: Malaysia, Indonesia and Brunei.

I crunch peanuts and put on a podcast about the war on drugs in the Philippines, where I'm headed next week to interview victims' families, just over the waves from here. You can see the islands of the southern Philippines from many parts of the eastern coastline of Sabah. It is a porous border that people shuttle back and forth across, many of whom are Suluk (also known as Tausūg), an ethnic group to which my biological grandfather, whom I never met, belonged.

The train gathers speed to a lazy clack. It seems the fight to get onto the train is going to be the most eventful thing about today's ride, as I've heard the first leg of this two-part journey is boring. As we pass the airport and a semi-new shopping complex, I think dreamily about the spectacular views I've been promised tomorrow on the second leg of the journey, from Beaufort to Tenom: the mighty churning Padas River, dramatic gorges and coffee plantations.

My nenek (grandmother) often talks about Beaufort as if it's a major world city (it has a population of about 13,000). Clearly, it is a repository for many of her memories of a nomadic life spent roaming from place to place in North Borneo. It took me a long time to realise that what she was saying is an English name – stupidly, never having looked at it on a map. I'd always heard it as 'Bofod', thinking it was an Indigenous name, until way later than I should have. The city was, in fact, named for Leicester Paul Beaufort – the British governor of North Borneo – in the late 1800s.

The train rolls alongside an interminably ugly highway which is interminably jammed. Sometimes, when stuck for hours in a jam, you'll hear sirens blaring and, painfully, cars will inch to the side to make way for what you think is an ambulance, but is in fact a rich man accompanied by a police escort, cutting a clean line through the traffic.

Sirens go off in my head regularly.

In the music of me, there is a predictable (if ragged) pattern; a sub-surface song cycle that has always been so. Since I was a teenager, my mood goes from periods of elevation – supremely confident, impulsive, intensely creative and often irritable – to extreme darkness, hopelessness and suicidal ideation. That is the overall pitch and heave of it, though within these periods there is erratic, staccato movement, where my mood can change within minutes. For two decades, I self-medicated with drugs and booze, which brought me to the gates of hell. For the past five years (thank god), the drugs and booze have gone, but the near-unbearable mood swings and cyclical depression remain. Earlier today, at the hospital, I received a diagnosis that's been a long time coming: bipolar disorder type two.

I wonder if this plays a part in why I have always embraced the speed and constant change of travel. The rocking rhythm of trains, the blurring zigzag of cars racing down highways, planes that cleave clouds, even the cadence of my feet as they jog along asphalt from A to B and back again – all have the effect of not just distracting but also tempering and modulating what I thought of as madness, but is, in fact, a diagnosable condition. Likewise, in the two decades since I started returning regularly as an adult (having only been twice as a child and once as a teenager), coming back to Borneo has always had a steadying effect on me. Taking the plunge and moving back here properly has been both illuminating and confounding: a ride simultane-ously into tomorrow and into the past.

The outer suburbs of Kota Kinabalu begin to give way to green fields and blue-tinted hills, along with glimpses of islands and the disputed South China Sea, often spoken of as a maritime frontier in the tussle between the US and China for global supremacy. Right now, the sun is tinting the sea rose-gold as it sinks rapidly through a rich green screen of trees. But, mostly, all I can see is highway.

*

Once upon a time, this lazy ride was named the North Borneo Railway and was the obsession of a man far, far away who built it for the sole purpose of exploiting natural resources and opening up the fertile interior of the west coast. North Borneo was under the control of a British trade company at the time, not the British crown itself. This company, British North Borneo Chartered Company, left behind sludgy mismanagement that was picked up with alacrity in the post-independence era, when British North Borneo became the state of Sabah and joined Sarawak, Singapore and Malaya in 1963 to form Malaysia.

There's always discussion about improving the railway line, but nothing seems to have changed in a long time, and its operation is erratic and lags far behind the service provided by trains in Peninsula Malaysia. In 2018, the then Sabah State Infrastructure Development minister Datuk Peter Anthony urged the Malaysian Anti-Corruption Commission to investigate the Sabah State Railway Department, calling the railway 'an embarrassment'. He cited myriad issues, from lack of training to the use of inappropriate materials for sleepers and oil container caps. He even said some of the materials were a hundred years old and should be 'sent to the museum'.[1]

One of the train doors suddenly swings open, with the train at full speed. Hot air rushes in, and the air-conditioning is rendered useless, but no one seems to mind. I go to use the bathroom and it's clean but the taps and flush don't work. Instead, the basin has six full plastic water bottles in it, with which you can wash your

---

1   https://www.nst.com.my/news/nation/2018/08/397157/sabahs-train-system-too-old-should-be-put-museum-instead

hands, and a full bucket of water to the flush the toilet sloshes on the ground. Maybe nothing much has changed in the seven years since the former minister made those comments (though he himself is now in jail, charged with falsifying documents). Outside, as the sun dissolves into sea, the ghostly outlines of the slatted wooden kampung houses and blue mosques with gold minarets slowly disappear into the darkness. Likewise, the massive elephant's ear plants and the rubber, papaya and banana trees. And also the neat rows and thick, uniform trunks and feathery fronds of the main cash crop that drives Sabah's economy – the oil palm – sometimes called 'green gold'.

The oil palm produces harvestable fruit three to four years after planting and remains extremely productive for twenty-five years. Palm oil is transported mostly by trucks and ships, and the major importers are China, India and the European Union. There has been big talk of creating a trans-Borneo railway, to allow easier and more durable transport of bulk cargo, including palm oil, petroleum oil and gas, timber and coal, between the Malaysian Borneo states, Brunei and Kalimantan, on the Indonesian side, especially if the national capital of Indonesia is moved there, as is planned. A feasibility study will be completed in August 2025. I wouldn't hold my breath on it being constructed any time soon, especially if the notoriously bad and tardily built Pan-Borneo Highway (which I've heard called the *Pain*-Borneo Highway on many occasions) is anything to go by. Likewise, there will be similar questions about the environmental impact on a fragile, diverse natural ecosystem and its effects on the lands of Indigenous communities.

But these questions feel like echoes of history. Before the cash crop of palm oil there were others, and these were the main reason for the construction of the railway.

In the late 1800s, tobacco farms were founded in the interior of Sabah, near Tenom. A Scottish man named William Cowie, through skulduggery, charisma and force of will, became the director of the British North Borneo Chartered Company. Cowie had been a close friend to Sultan Jamal-ul Azam of Sulu and was one of the reasons why there was a British presence in Borneo in the first place, having set up a trade company, the Labuan Trading Company, to evade Spanish blockades to smuggle opium, tobacco and weapons to Sulu. His closeness to the sultan allowed him to help Alfred Dent, a British merchant, acquire a concession treaty to much of North Borneo for 5,000 dollars. Dent went on to found the British North Borneo Chartered Company in 1881. As plots of land were sold to British planters deep in the jungle, and rubber, cocoa and tobacco plantations began to be established, Cowie became obsessed with the idea of building a train line from Jesselton (modern-day Kota Kinabalu, where I joined the train) to Tenom.

Whereas previously the British North Borneo Chartered Company had been run by governors based in Borneo, when Cowie was elected to the board of directors in 1884, he decided to pull the strings from London instead. From the lavish dining rooms of the Criterion Restaurant in central London, Cowie and his band of investors decided the fate of North Borneo and its inhabitants. As historian Ross Ibbotson says in his book *The Building of the North Borneo Railway and the Founding of Jesselton*, Cowie acted 'as if he was the sole authority on North Borneo, personally familiar and experienced with all aspects of the territory. In reality, apart from his visit in 1898, he had not set foot in the interior.'[2]

---

2   Ross Ibbotson, *The Building of the North Borneo Railway and the Founding of Jesselton*, Opus Publications (2018), p. 82, fn. 30

Cowie's plans were heavily criticised, especially by Alfred Dent, the founder of the company, and his supporters, who saw them as a waste of money and who eventually resigned from the company over this issue. Ibbotson says:

> ...until his death in September 1910, Cowie dictated the policy of the British North Borneo Company with complete disregard to the opinions of his peers or the senior staff in the field in Borneo. In his mind the development of the railway took precedence over anything else and he allowed no one to criticise this policy.[3]

The company purported to have two responsibilities in North Borneo: 1) economic development through exploitation of natural resources and 2) protecting the rights of the inhabitants. However, despite grand pronouncements about ending slavery in the region, the former responsibility completely undercut the latter. The company's intention to cut a swathe through the jungle for the purposes of resource exploitation – to tame and hack and dig – required the exploitation of local Indigenous labour, as well as the importation of additional Chinese labour (a huge death toll has been recorded), along with Sikh officers to quell local opposition.

It still trips me out to look at the photos of these white men, with their big white moustaches and heavy suits, and think about the sway they held over the past and future of this land. How incongruous they seem in the landscape, almost laughable, if their effect hadn't been so brutal. Of course, it cannot help but make me think of Australia, my childhood home, and

---

3   Ibid, p. 32

the British soldiers in their heavy blood-red jackets, marching through a landscape they would stain the same red.

Where once these white men in heavy suits rolled towards the riches of the interior, now the train is full of locals. There is just one lone white traveller on the train – a tall man with a backpack (I have no idea where he comes from, though to my eyes he looks Mediterranean). He is stared at, an oddity. Most tourists to Sabah come from China and Korea. Europeans and Australians mostly come for the wildlife, but if it's out of historical interest, it's usually related to the Second World War, in particular the brutal Japanese 'death marches' from Ranau to Sandakan (my dad's home town) that killed 2,434 Allied soldiers. The North Borneo train network that so obsessed Cowie was mostly destroyed during the Second World War, as North Borneo became a theatre of war. Many locals were also killed during this time, not just by the Japanese but in Allied bombing attacks on local towns, purportedly to smoke out Japanese soldiers, but with plenty of collateral damage. Sabahan historian Avtar Singh tells me that there is no data on the number of people killed, and it is just based on word-of-mouth interviews. 'Local death toll figures vary between 2,000–6,000,' he tells me in a WhatsApp message, but also cautions that 'Not all deaths were related to war.' As well as Japanese intervention and Allied bombings were causes such as poor nutrition, old age, death during childbirth, poverty, lack of medical services, etc.

Even though my skin is brown, locals always stare at me on the train too, categorising me as a tourist, like the tall backpacker. In a way, of course, I am. I have often described myself as a 'wayfarer in the homeland', never more reminded of my Australianness than when I'm here in my tanah air (homeland). My mother's ancestors come from a different green island – Ireland, the Emerald Isle – but that's a story for a different essay. Often, locals will look at my beard and presume me to be Arab or Pakistani. From daily experience, I know what my fellow travellers' response will be when they hear that I speak Malay and have Sabahan heritage – shock, pride, excitement. Yet, still, my brown skin and my heritage will never change the fact that my dress, my gait, my speech will always betray me as an orang putih – *a white man*.

I once heard a story about a black guy from the UK visiting a tattoo shop here, and when he left, someone asked:

'Who was that orang putih?'

'Eh, which one?'

'That person that just left!'

'You mean that black guy?'

'Ya, that orang putih that just left!'

Damn. I guess race really is just a construct.

Beaufort (or 'Bofod') is now an hour away.

Beaufort and Sipitang are the main Sabahan towns where my grandmother's ethnic group – the Kadayan – reside, living alongside the Murut, Dusun, Lun Dayeh, Bisaya and a Hakka population originally brought over to work on the train line. She is from a small village called Mesapol. There are also large populations of Kadayan in Brunei, Labuan and across the border in the state of Sarawak. The origins of the Kadayan ethnic group are regularly described using words like 'unclear', 'puzzling' and 'confusing'. While they are considered a Dayak (Indigenous) group, an often-repeated myth says that they originally came to Borneo as a retinue of helpers (or guards, or even slaves) from Java to help teach the Sultan Bolkiah of Brunei how to cultivate wet rice. Amde Sidik, a Kadayan academic, rejects this theory, saying in his book *The Mystic of Borneo: Kadayan* that there is evidence of their presence here from long before that time. To further complicate the matter, Wikipedia (albeit with citation needed) tells me the Kadayan language shares a 90 per cent similarity with the Banjarese language in Kutai, East Kalimantan (on the Indonesian side), which is a very distant part of Borneo. I've even heard tell of the Kadayan being related to the Kelabit people in the highlands of Sarawak, or to the Kadazan people, since the names are so similar. It's a rabbit warren that makes me feel exhilarated or confused or hurt and dislocated, depending on the day.

My grandmother is in her nineties (we think) and may not be with us much longer, and is also possibly my closest connection to understanding some of these things. But she, through her own hurt and many superstitions, is often unwilling to pass on information, or seizes up when she knows she's being recorded. In my many writings about her, I always refer to her simply as Nenek, a mythic person, painted in simplistic strokes, but I have, I only realise now, always left her nameless. So, for the record, let me name her here in full: her name is Lauya Binti Garai.

The train continues into the dark.

When I arrive in Beaufort, the town is quiet and already shutting down, at 8.30 p.m.

I find a bright Mamak shop and order murtabak ayam and an iced teh tarik, continuing to listen to my audiobook about the war on drugs in the Philippines. My phone lights up. The homestay I've booked for the night messages to ask when I'm arriving.

Be there soon! I reply. I'll find a taxi.

There are no taxis at night, sir, the homestay replies.

Ah, OK. So is there any way you can pick me up? I ask.

No sir, I'm sorry, we cannot.

(I look at Google Maps. It's a fifteen-minute car ride, but a three-hour walk.)

OK then… So… what do I do?

Well, replies the homestay, maybe you could cancel this booking and book a hotel in town.

Great… I say. Thanks a lot.

An hour later, still fuming, I check into the River Park Hotel. I set my alarm for 6 a.m. and eat some M&Ms, soothing my soul with thoughts of the adventure that is to come tomorrow morning, when the train carries on its journey, through valleys and coffee plantations, alongside the churning brown Padas River, towards Tenom. My friend Suffian's main memories of taking the train as a kid in the 1970s are of the wooden seats and the windows wide open, and that at every stop, hawkers selling delicious food would leap onto the train, plying kuih muih, mee goreng, boiled eggs and drinks from rattan baskets covered with banana leaves. His own father-in-law remembers that back in the fifties, before the logging boom, the waters of the Padas River were not brown but turquoise.

As I look out the window, I can see the river glinting vaguely. Nenek was born so close to here, sometime in the 1930s, when the train was still used to exploit the interior. I wonder what stories she has about it. She'd surely have some. I want to ask her about it, but sadly, a hearty dose of family melodrama and her lack of a mobile phone has made her uncontactable. Also, of late, Nenek seems to be dissolving into dementia, rattling, like this train, into the dank, shapeless dark of memory. She shuttles backwards and forwards from the now into the deep past, a past inaccessible to anyone else, a past in which somebody tried to poison her when she was living on the street, which she now accuses everyone in the family of doing.

I think about Nenek when she was younger – the various traumas and ingenuities (particularly improvised poetry) that made her. Conditional factors – landscape, bad men, local herbal remedies, stories and superstitions passed down, prayers. Then those other things – mood swings, irritability, obsessive-ness, impulsiveness – that are so often shaped and distilled by conditions but are most likely in-built in the blood. These things that my father and I also share. It isn't my place to diagnose somebody else, but the psychiatrist at the hospital did seem keen to interrogate my family's mental health history and mentioned that bipolar is often hereditary – in 60 per cent of cases.

I am in a constant tussle with myself about writing in this way about Nenek. Hell – about myself. Always digging deeper and deeper. Mining. Extracting. Just as the British exploited the interior, am I not doing the same? Telling stories not mine to tell, varnishing them like Borneo ironwood to furnish someone else's leisure? Yet her story is my story. It lives in the churn of my blood and the loam of my brain. It often feels like I can't tell mine without telling hers. Seamus Heaney said his pen was a spade. I worry I dig too deep.

Nenek, in the past, munching a bucket of her favourite KFC drumsticks, told the family I have severe mental health problems. She might point to spirits or angin, wind that gets into your body, almost like the Shakespearean humours. My dad might point to Shaitan – Satan. I imagine my dad and Nenek scoffing at a psychiatric diagnosis: the suggested solution to all that ails me has always been that I should pray or get married. I am now indeed happily married, and I do have my own very private religious practice, but the mood swings and cyclical depression remain, so... so much for that theory.

In the pitch black night, bright stars are visible, a galaxy of antipsychotics.

*

I wake up at 6 a.m. and look out the window of the River Park Hotel. The Padas River is moving past swiftly, brown, and the sky azure above. I'm so excited for the next part of the ride, onwards to Tenom, where locals hang out of the moving carriages, and the valleys fall away crinkled and redolent. A place where, I've heard, the local Murut people cut down the electricity lines to melt them down for bullets to shoot at the invading Japanese troops in the Second World War.

I ask the friendly young fella in the hotel's reception when the next train to Tenom is, and he cheerily replies that there are no trains to Tenom due to a landslide. Landslides are common here in Borneo on the train line and the roads, due to heavy rain, soil erosion and poor construction/lack of slope stabilisation. The train line has been out for a while, he tells me, and won't be operational for the foreseeable future. Nowhere had I seen on the internet that there had been a landslide, but that could have been because I stubbornly insisted on not using Google Translate to interpret the notices on the Sabah Railway website.

Reading up on how train lines around the world deal with landslides, from motion sensors to vibration-detecting fibre-optic cables and slope stabilisation, I smile wryly. I can't imagine any of these things in Sabah. I think of something a friend once said to me: in Singapore, everything works perfectly but it has no soul. In Malaysia, nothing works, but at least it has soul. I'm sure there are a million versions of this adage around the world.

So there's nothing for me to do but drink Tenom coffee (the closest I'll get to the town itself), buy some delicious pandan cakes for my wife, and then wait for the train back to Kota Kinabalu. This time, what was mostly in blackness will be in full sunlight.

As the train leaves Beaufort for Kota Kinabalu, passengers see massive, extravagant Chinese temples, a nod to the early Hakka labouring communities. I read on the MySabah travel website on the estates along the train line, the mortality rate of Chinese labouring communities was often 40 per cent due to poor treatment and living environment.[4] Passing through Membakut you see a large wooden building that served as a train station back in the day, the ground floor of which is occupied by Chinese storefronts. We roll on. Padi fields, oil palm plantations, the same boring, broken highway, all under an unrelenting sun. Finally, we pass Kinarut, possibly named thus as a corruption of 'China Road' (this is disputed, some say it is a Dusun word), yet another nod to the early labouring communities. As we approach my new home of Kota Kinabalu, malls and dangerously lowered cars appear, belching smoke. Home, yes. Some of the journeys that change us are dramatic, eventful. Others, like this one, are seemingly banal, a simple ride from A to B and back to A again. Funny how a train ride, where the most anticipated, exciting part is missing, can still be so illuminating. The train line is still working – a greedy, powerful man's freight train transmogrified into a muted myth, truncated and fragile. Trains, trains, carrying our souls.

I travel to the interior of myself.

---

4   https://www.mysabah.com/wordpress/north-borneo-train-tour/

The Cascade Mountains, as seen from the Empire Builder train.

HOW TO NOT HACK YOUR TRAVEL
SHAHNAZ HABIB

The Mississippi River is a frozen shadow outside the window when we sit down to dinner; four women who don't know each other. Audrey is a tall, elegant surgeon in Fargo, North Dakota. Julia is a fashion designer for Target in St Paul, Minneapolis. Ann works at her friend's hardware store in Starbuck, a tiny town in Wisconsin that is difficult to find on Google Maps, because it thinks I am looking for a Starbucks in Wisconsin. Ann is eager to retire but her friends need her and she can't leave them in the lurch because there's no one in the town who can take over her job. So what can she do? She is stuck working when she would rather spend time with grandchildren. Although she will never again go to the waterpark with five teens. That was exhausting. Then there is me – a writer from New York. A writer! Are you going to make us characters in a novel? Good, because I want a more interesting life. Ann provides the small-talk glue we need as we come together awkwardly – four women in the dining car of the Empire Builder.

Let's pause right there, before our movable feast arrives on the white tablecloth-laid dining table – rigatoni with plant-based

Bolognese sauce for me, steak for Audrey, Atlantic salmon for Ann and Julia.

It's quite the title – the Empire Builder. None of the other trains that ply the fat body of the United States bear such a grandiose name. There is a poetic twang to the Crescent (New York to New Orleans and back). The Coast Starlight (up and down the West coast) is named like a sonata while Lakeshore Limited and Capitol Limited (both Washington DC-bound, from Chicago and New York) sound busy, they have things to do, places to be. Even the Texas Eagle (Chicago to Los Angeles and back) carries its patriotism subtly, on the wing of the national bird, and a quick visit to the dictionary reveals that California Zephyr (cutting through the Rocky Mountains to connect Chicago with California) and Palmetto (bridging New York and the southern states) are meant to evoke sweet breezes. The Floridian is a temporary train that no one has even bothered to name properly. I could go on and on. Sunset Limited, Silver Meteor, Cardinal – none of these names make you cringe. And then there's the Empire Builder, which runs from Chicago to the Northwest, its name and route proudly implying the westward expansion of the United States, the genocide of the Indigenous people of North America, the myth of white Christian manifest destiny.

In our era of crowded airports and billionaires' space flights and air traffic controller strikes, trains have excellent PR. They represent a slower, more luxurious way to travel, a kindness to yourself and the world. Nobody is weighing your bags, the seats won't crush your knees. The glorious ineffi-ciency of a long-distance train journey comes with a sepia mist of nostalgia. But in the nineteenth century when the first railroads were being built, they were not meant for gentle travel. The Liverpool and Manchester Railway, the first inter-city train route in the world, was built to carry raw cotton from the port of Liverpool to the factories of Manchester. Cotton was central to colonialism. One of the reasons the United Kingdom became the empire it became was cotton, plucked cheap in India and then returned at a huge markup as finished fabric, slowly strangling India's indigenous textile industries.

In the United States, the first railroads transported not only goods but also settlers. The early railroad companies were not just in the transportation business, they were real-estate developers; they got land grants from the United States government, which allowed them to lay down tracks westward, displacing the Indigenous people who lived in those regions and stewarded those lands. The US army enforced construction, massacring entire villages. In turn, the railroad companies sold the land to settlers and then transported those settlers to those lands. This is the genius of settler colonialism – the customer is also the product.

The Empire Builder was inaugurated in 1929, long after the West was colonised. But the name is a homage to James Hill, a railroad director who built the first transcontinental railway route, the Great Northern Railway, by consolidating multiple railroad companies. Hill travelled by horse up and down the West, identifying the best routes, before laying down the tracks for the iron horse. Along those tracks new towns emerged, and soon timber and agriculture and mining industries followed. The railroad disrupted the rhythms of Indigenous life, trade and land stewardship. Stubborn, arrogant, hardworking – James Hill was one of those template capitalists in whose image the United States was cast.

There's something else that is named after James Hill, a far more appropriate legacy. Jim Hill mustard (Latin: *Sisymbrium altissimum*) is a mustard bush that was once indigenous to the Mediterranean. Legend has it that some seeds hitched a ride with another crop that was travelling on a James Hill train, blew off the wagon and flourished along the sides of the rails. The Eurasian mustard bush quickly took to the American plains. When mature, the plant uproots itself and turns into tumble-weed, spreading its seed. A single plant putting out delicate yellow flowers one spring can produce an infestation by the next year. Jim Hill mustard is a pest, a weed, an invasive species.

One-and-half centuries after James Hill bought his first railroad, I am sitting in the Empire Builder's swaying dining car, eating my vegan Bolognese with my new friends, while the train rattles through the plains of North Dakota. It is 1 January, which makes Audrey, Ann, Julia and me four of the very first of the 387,953 passengers who will travel on the Empire Builder in 2024. Audrey is a regular on this train. This time she is going back to Fargo after spending Christmas with her fiancé and family – he was boyfriend before Christmas, she tells us as she raises her left hand so the ring can glint for us. Julia's sister had a baby, so the family went to her house for Christmas. Otherwise, she would be driving right now and stuck in traffic somewhere, she says ruefully. Ann was visiting grandchildren. 'I always take the train, it's how I unwind after all the holiday stress,' she tells us.

'And you, hun?' Audrey is gracious, with perfect bedside manners, even when not required. I want her to be my surgeon. I am on my way to a book reading in Seattle. I started in Washington DC, where I was spending the holiday with in-laws, I tell them. A brief layover in Chicago to switch trains, now onward to Seattle where I will be reading from my newly published book about travel.

Except my layover was not brief. I don't tell them this, because I am too embarrassed.

*

My train from Washington DC (the indifferently named Capitol Limited) arrived in Chicago the day before, a little after 10 a.m. The Empire Builder leaves Chicago for the northwest in the afternoon. I checked Google Calendar – the layover was multiple hours. How often does one get to have a lunch break in Chicago? My husband, who studied acting in Chicago only to find out that he was terrible at it, never wants to return to this city. This, then, was my chance to see Chicago. For free! I felt somewhat giddy at the thought of getting a little trip within a trip. It was like something that would be in one of those online travel hack discussions where I am a constant lurker.

I will admit – I love a travel hack. I have the credit card that also gives me travel insurance and points. My inbox is a jumble of mistake fare alerts. I am committed to the shoulder season, even while climate change has made the term irrelevant. Much of my free time is spent searching 'Anywhere' on Google Flights. Of course, I buy local SIM cards when I arrive in a new country. Obviously, I wear my heaviest coat and shoes on the plane to free up luggage weight. And yes, I have all the airline apps and I am practically on first-name terms with the bots who live inside them. I always ask to pay in local currency when I travel, never in dollars. And while I have never actually skiplagged, I stand ready to skip and lag should the opportunity arise. In short, I don't let a single travel hack pass me by without a respectful tip of the hat or a screenshot.

So, of course, I travel-hacked my lunch break in Chicago. I downloaded the public transportation app. One of my credit cards gives me access to a museum membership through which I had a free ticket to the Arts Institute of Chicago. I mapped and timed my itinerary – lunch, bookstore, museum. A perfect little bite of Chicago.

And it was. Chicago was rainy but I had an umbrella. (Travel hack: folding umbrella at the bottom of your backpack, never take it out, except, obviously, when it rains; put it back as soon as it dries.) The restaurant was busy but I shared a table with another diner. (Travel hack: ask fellow single diner if they are OK with sharing a table.) Unfortunately, the bookstore did not carry my newly hatched book, but this did not dampen my enthusiasm at all. (Life hack: remember that you are not a celebrity.) I took the bus to the museum where my ticket, booked for free via the app, also let me circumvent a long line. It's a series of travel-hack wins.

Inside, the museum was soft and calm. My favourite exhibit was a video by Marwa Arsanios. *Reverse Shot* was speculative and experimental: a privately owned quarry in northern Lebanon wants to become common land. Can land have free will? Can land shape its own manifest destiny? Acting in the quarry's name, a group of activists seek to file paperwork that will turn the land back into a trust (*waqf*) or public domain (*masha'a*). They sit on the white rocks in the quarry and brainstorm.

The piles of white rock in the Lebanese quarry, left behind after mining, reminded me of a photo I saw when researching the history of trains in the United States for my book. A mountain of white bison skulls. As railroads started encroaching the West, trains started bringing hunting parties that would take aim at herds of bison from the windows. It was impossible to miss because there were so many bison. Millions of them once roamed the Western plains – they were the largest land mammals on the continent. The mass slaughter was not simply recreational; it also helped decimate the Native Americans who depended on the bison population for meat and hide. By the time the photo was taken in Michigan in 1892, a population of millions of wild bison had shrunk to 456.

An old farmer, interviewed in *Reverse Shot*, remembers how, long ago, the quarry used to belong to whoever grew food in it. My feet felt light on the ground as I caught a glimpse of a world in which land belongs to no one and to everyone. But then it was time to return to Chicago Union Station and board the Empire Builder.

I know what you are thinking. That I lost track of time and forgot to board the train. In fact, I did not lose track of time. Even while watching the video in a dark room, I kept glancing at my phone. I wanted to get back to the station at least an hour before departure. (Travel hack: pilfer snacks from the lounge.) I had set a calendar alert and at 2.30 p.m., my Google Calendar, which had automatically generated a calendar event for my train reservation, sent me a notification – it was one-and-a-half hours before the Empire Builder was due to leave Chicago.

Outside the museum, a protest caravan for Palestine made its way down Michigan Avenue. It was the third month of the genocide in Gaza, and there were protests everywhere all the time. Writing this now, more than one-and-a-half years later, those early days when we did not know how long the massacre would continue seem to belong to a faraway time. Any minute now, we thought, there would be a ceasefire.

Since the bus back to the station was delayed by the protest caravan, I walked. It was a twenty-minute walk and I arrived at the station a few minutes after 3 p.m., with most of the hour to spare before the train left. I grabbed a chocolate bar and a packet of crisps from the lounge and took my luggage ticket to the front desk so that I could retrieve it from storage.

'Where were you??? Your train just left!' The woman behind the counter told me.

When Google picked up my train reservation from my Gmail and inserted it into Google Calendar, it had automatically converted the 3 p.m. departure into my local time zone, where it is now 4 p.m. I had relied on the calendar reservation, set my alarm based on it. Just as I got to the station after my leisurely walk, my train pulled out of it. The train to Seattle, a two-day ride across the Midwest and the Great Plains, a dream trip that I had saved money and time for – it was gone.

The horror I felt when I realised this was only second to the humiliation. I was a forty-five-year-old middle-aged woman. I had just published a book about travel. How could I do something so stupid as miss a train that I had a seven-hour layover for? There were no train delays, no bad weather, no accidents. Just unadulterated stupidity bottled fresh at the source.

An hour later, I had a ticket on the next train to Seattle, purchased for almost double the price I had paid for the train I missed. Luckily, the Empire Builder is a daily train and luckily the train next day had roomettes available. I have never used the word 'luckily' to describe a situation in which I spent so much money unnecessarily, but there it is. The biggest 'luckily' of course is that I had the money for the second train ticket. This would not always have been the situation, but after a few years of steady employment, at least the money existed. So I bought the ticket and decided to immediately repress any thoughts about the amount of money I was wasting. Of course, what this meant was that I had two constant thoughts in the back of my mind: money, money, money, don't think, don't think, don't think, each thought a metronome against which the other banged itself.

Now I needed a place to sleep. Did I mention that it was New Year's Eve? Naturally, the hotels were all booked. As revellers gathered along Chicago's Magnificent Mile, I sat in the train station calling one hotel after another. Only the most expensive ones had any room availability. Soon, I was standing in the opulent lobby of a luxury hotel that had one room left. Money, money, money, don't think, don't think, don't think.

'And can I have your ID please?' the receptionist asked me.

'I don't have an ID,' I admitted. Yes, I had left home without an ID. Since the trains do not ask to see ID when boarding, I hadn't thought to bring my passport, the only state-issued ID I have.

'I am sorry ma'am, but I need to see an ID.'

There's a book with a whole chapter about how passports went from permission slips for travel to documents that states use to control identity and mobility. I wrote that book. That was the book I was going to Seattle to do a reading for. If anyone should have known better than to travel without a passport, it was me. Yet, that was exactly what I had done.

The hotel manager explained that he was obliged to check IDs for the safety of his guests, to protect them from identify fraud and terrorism. I wanted to tell him that this has nothing to do with safety. It's simply a small act in the enormous security theatre we all perform in. In fact, a few years ago, hackers – probably state-sponsored from China – infiltrated the database of the hotel chain that owns this luxury hotel. It is the largest hotel chain in the world. The hackers managed to access the personal information, including passport numbers and dates of birth, of about 500 million customers. Handing over your passport to a hotel reception desk is not the safety flex that this manager was mansplaining to me.

But I did not tell him this, of course. Instead, I thew myself on his mercy. I told him my sob story. I showed him my insurance card, my credit card, a digital scan of my passport from my email. A damsel in document distress. Other customers checking in gave me pitying glances.

When he finally relented, it was after I showed him my book jacket. My photo on the inside flap, right above my bio. After an overnight train journey, I did not look like the merrily laughing woman in that photo. Nevertheless, he was convinced. What kind of terrorist or identity thief would write a book and walk around with it as ID instead of just buying a fake passport on the dark web. For one night, I am the owner of two hundred expensive Chicago square feet. Money, money, money, don't think, don't think, don't think.

The next day at dinner with Ann, Julia and Audrey, I do not tell them all this. I omit my miserable evening in Chicago – bursting into tears when I called my husband from the hotel, waking up from my exhausted sleep at midnight at the sound of fireworks. As the train leaves Wisconsin behind and crawls into Minnesota, we wish each other goodnight. Perhaps they too have secrets they do not confess. Maybe Audrey is marrying for money, not love. Maybe Julia cannot stand her sister. Maybe Ann has considered setting fire to her friend's store. These are the best stories: that ones we do not tell.

*

The sun rises over North Dakota as I wake in my tiny train room. After the urban sprawl of Chicago and Milwaukee and Minneapolis, and the bright lights gleaming over the Mississippi and Wisconsin rivers, North Dakota is a vast field of snow. Freight trains pass us often and occasionally there are warehouses or some complicated mining infrastructure to be seen, but little by way of houses and humans.

At Minot, where the train stops for a full forty minutes to refuel, I step out and stretch my legs. Some brave souls are venturing beyond the station to explore the town, but my desire to cram in extra journeys during layovers is at an all-time low. Also, I am freezing, even in my sleeping bag of a winter jacket. There's an Indian father and his tween daughter walking the platform. We immediately congregate, the only three brown people. Along with the various African American train staff, we have created a brief spike of diversity in this town. He is a train enthusiast and we start trading stories of Indian trains. I tell him about the time a stranger on the Kerala Express gave me his blanket on a freezing cold night. He tells me about the time the Mangala Express had to spend a whole day unscheduled in a coastal village because it was monsoon and there were landslides ahead. As a young college student in Delhi, I was under parental instruction to travel by the Kerala Express, never the Mangala, because of the landslides. So, of course, I would forget to book my tickets until the Kerala Express no longer had any seats available. The Mangala curved down the west coast of India instead of cutting through its heartland, running alongside the ocean at times, or passing one waterfall after another. Even the snacks were more delicious on the Mangala.

As the two of us reminisce about pav bhaji and potato chips, I fall through a memory hole. The cold winter morning on the American prairie turns into a rain-washed Konkan coastal village at dawn. When the conductor calls 'All aboard', for a moment I am in two places at once, a foot on each continent.

At lunchtime, my companion is a retired geologist. He is pleasant and smart – at first. Then he starts talking of the people who worked under him in North Dakota. He shakes his head and says, 'I could never understand them. They never wanted to leave this wasteland.' The word catches in my throat. 'There is nothing here,' he says gesturing at the plains we are speeding through. 'But this is where they wanted to be.' He is mystified. Scott is retired from the second largest oil service company in the world. The company he worked for has been conducting massive fracking operations in the north-west. He continues to consult for them.

A few minutes later, we are in Montana and he is denying climate change. I am alarmed but also thrilled. I have never met a flesh-and-blood climate-change denier in person. I listen eagerly. Scott is not a garden-variety conspiracy theorist. As a scientist, he has managed to come up with a sophisticated bubble of denial. The oceans are warming, he admits, but this is because of solar pulsing. We were apparently living through an Ice Age and now we are not. It's all perfectly normal. And fracking is perfectly safe.

This should be my exit cue, but I can't bear to step away. I have to admit, I love what he is saying. It is fiction, but such a delicious fiction. I want it to be true. What must it feel like to move through the world without any environmental guilt or climate anxiety? Who would I be if I could just hit delete on all the worrying I do about the world my daughter will inherit. My feet feel light on the ground as I catch sight of a world in which we did not mess up the delicate equilibrium of heat, light, air, water and earth.

Soon we arrive at the most scenic part of the westward journey on the Empire Builder: the southern edge of Glacier National Park. There is a frisson of excitement all over the train. The observation car has become crowded as passengers line up to sit along the windows. But it is a winter evening and the sun has set and the glaciers are barely visible. Using Google Maps, I identify a distant, dark hulk to be Summit Mountain but then spend the next twenty minutes second-guessing myself.

The next morning makes up for this. When I wake, it is in a winter wonderland. We are in Washington now, having entered and left Idaho in the night. The train is rolling through the foothills of the Cascades and its craggy mountains are dusted with snow. Fog rises amidst mountain peaks, and streams gleam like crystal necklaces. Most of the landscape is in the sepia tones of north-western winter, but suddenly a jade green river will push through. There are miles and miles of forest. Even the train has slowed down, so it feels like we are travelling in slow-motion.

Riding the long-distance trains in India is how I learned my country's geography. I learned about the great rivers of India – Ganga, Yamuna, Narmada – in school, but until I heard the long, deep rumble the train makes when crossing the Narmada, what did I truly know of their span and depth? The Western Ghat mountains were an abstraction until I rode the Sahyadri Express. Deep inside the mountain, in the darkness of the tunnel, I realised what it was to have a mountain rise like a wall, how it cuts off people and clouds and climates.

And now this train ride across the United States is giving me a feel for the landscape of my adopted country. Even when

I'm simply staring out the window, lost in my own thoughts, I'm taking in its vastness and variety. The great flat sparseness of North Dakota was only yesterday and now here I am in the thick of lush forests. Sometimes in a Feldenkrais lesson, my teacher will have us start by considering the map of our body, the shape it makes as it lies on the ground. Then, over the course of the lesson, as we tune in to what's going on in each part of the body, the map deepens and fills in with detail. My ride on the Empire Builder feels similar: a body scan of the United States. The map in my head, which used to be a Google Map, zoomed out from Brooklyn, New York, is now richer, still incomplete of course, but much more detailed – it has colour and movement and smell and feelings.

When they say train rides are scenic, what they mean is iconic sights, such as the glaciers or the Cascades – these stunning landscapes that bring your heart to your mouth and remind you how lucky you are to see this. But even in this train route that is famed for scenery, the gorgeousness is a very small proportion of the ride. Granted, I am travelling in winter, but even at the best of times, a two-day train ride is guaranteed to have hours and hours of boredom and nothing to see, punctuated by a few minutes of awe-inspiring scenery. Watching the Cascades cascade away outside my window, it occurs to me that we do not enjoy scenery because it is scenic; we enjoy it because we have been bored. That is what the train asks of you: a willingness to be bored. And perhaps that's why some of us cannot get enough of trains. It rearranges our relationship with time. It is a few hours later that the train curves around the Seattle Sound, arriving at its final destination, but it feels like minutes.

Time is slippery. You can keep checking time without realising what time it is. You can slide from one moment into another, making terms such as past and present feel meaning-less. You can hope for a ceasefire while knowing that the fire will not cease, that history will keep looping. Like the earth, time cannot be owned. Yet we cannot stop trying.

In Montana, there is a small town called Havre. A boom town from another century. The Empire Builder stops there for about twenty minutes. In front of the station, looking out at the mountains, is a statue of James Hill. He looks a bit tired, somewhat unloved, Ozymandias in Montana. The copper is turning blue. Today's empire is tomorrow's odd curiosity.

*

I, too, feel humbled by my days in the train. Dear reader, can you feel an epiphany coming? This is how travel essays are supposed to end, and I loathe it. The travel epiphany – the pretty piece of wisdom that mystically alights on the writer of a travel story. You probably detest it, too, but unfortunately, I have an epiphany for you. And I don't know how to say this without sounding stupid or obvious, but here goes. The ultimate travel hack is to get on the train.

Board it before it leaves. You don't have to compress your luggage or find the cheapest flight or explore five places in one trip. You don't have to be invited to a charming local wedding. You don't have to have the Most Exciting Trip. As long as you manage to board your transportation, as long as you move from one place to another, you have hacked it. You have hacked time itself.

ABOUT THE AUTHORS

Jack Curtis is the co-founder of Carbon Jacked, an environmental start-up that helps businesses and universities get the people part of sustainability right. Prior to that, he worked for HM Treasury as a senior policy adviser and trade negotiator. He also has experience in business, policy and campaigning.

Shahnaz Habib is a writer and translator based in Brooklyn. She translates from her mother tongue, the south Indian language of Malayalam, and has translated two novels by Benyamin, *Jasmine Days*, winner of the 2018 JCB Prize, and *Al Arabian Novel Factory*. She is the author of *Airplane Mode: Travels in the Ruins of Tourism*, winner of the 2024 New American Voices Award.

Clare Hammond is a British journalist. Based in London, she works for non-profit Global Witness, investigating issues relating to natural resources, conflict and corruption. In Yangon, where she lived for six years, Hammond was most recently the digital editor of *Frontier Myanmar*, an investigative magazine. Her first book *On the Shadow Tracks: A Journey through Occupied Myanmar* was published in 2024.

Leon McCarron is a writer, broadcaster and hiking trail designer from the north of Ireland. He is a Yale World Fellow and the recipient of the Cherry Kearton Medal from the Royal Geographical Society. Leon is the author of three books. His fourth, on the Hejaz Railway, will be published in 2026.

Andrew Martin's many books include the Jim Stringer series of historical railway thrillers and half a dozen railway non-fiction books, of which the most recent is *To the Sea by Train*. His latest novels are *The Moquette Mystery* and (as A.J. Martin) *The Night in Venice*. He is also the author of the Substack Reading on Trains. His website is at martinesque.co.uk.

Omar Musa is an author, visual artist and poet from Queanbeyan, Australia. He has written two novels (including *Fierceland*), three books of poetry, five hip-hop records and two acclaimed plays, *Since Ali Died* and *The Offering*. His work has appeared in *The Best Australian Stories* and *Best of Australian Poems*. His debut novel *Here Come the Dogs* was long-listed for the International Dublin Literary Award and Miles Franklin Award and he was named one of the *Sydney Morning Herald*'s Young Novelists of the Year in 2015. He has had several solo exhibitions of his woodcuts, including his most recent collection All My Memories Are Mistranslations. He is based between Borneo and Brooklyn.

Mark Ovenden is an author, lecturer and television/radio presenter. His work focuses on the design, architecture, cartography, signage and typefaces in the world of transport. His first book *Metro Maps of The World*, published in 2003, was followed by one about the Paris Metro. Subsequently he has published books on the subjects of railway maps, London Underground design, Manchester Metrolink, typefaces, airline maps and underground cities. Mark is a Fellow of the Royal Geographical Society and in 2024 graduated from the University of York with a masters in Railway Studies. He was born and brought up in London but also lived in various French, American and UK cities. He is currently based on the Isle of Wight where he hosts a weekly radio show and podcast.

Yvonne A. Owuor was born in Kenya. She is the author of *Weight of Whispers, Dust,* which was shortlisted for the Folio Prize, and *The Butterfly Sea.* Winner of the Caine Prize for African Writing (2003), she has twice received an Iowa International Writers Fellowship and was shortlisted for the FT/OppenheimerFunds Emerging Voices award. Her work has appeared in *McSweeney's, Granta*'s 'The Politics of Feeling' and other publications, and she has twice been a TEDx speaker (Nairobi and Euston). She has been a resident and fellow in several places including the Lannan Foundation, the Wissenschaftskolleg zu Berlin, Civitella Ranieri, Dorothea Schlegel and the Stellenbosch Institute for Advanced Studies. She spends most of her time in Nairobi, Kenya.

Vicki Pipe has been researching and writing about railways ever since she co-founded, produced and presented *All The Stations,* an online documentary project visiting every railway station in the UK and Ireland. Her key interest in the railways lies with the stories of people, social change and how the railways influence our sense of space and surroundings. She is the co-author of *The Railway Adventures: Places, Trains, People and Stations,* and lead author of *Great British Railways: 50 Things*

To See and Do. She is a former columnist for *Modern Railways* magazine, where she focused on the work of Community Rail and its impact on the lives of people around the country. Outside of writing, Vicki is a museum and heritage engagement specialist with many years' experience producing cutting-edge programmes, activities and exhibitions with a diverse range of audiences including adults, families and young people. She has worked for organisations including The Royal Collection, London Transport Museum, Bow Street Museum of Crime and Justice, Imperial War Museum, The National Trust and Bletchley Park.

Monisha Rajesh is an author and journalist whose writing has appeared in *Time* magazine, *Vanity Fair*, the *Financial Times* and *Travel + Leisure*. In 2023 she was named in the Condé Nast Traveller Women Who Travel Power List. Her first book, *Around India in 80 Trains* (2012), was one of the *Independent*'s top ten books on India. Her second book, *Around the World in 80 Trains* (2019), won the National Geographic Traveller Award for Best Travel Book and was shortlisted for the Stanford Dolman Award. Her third book, *Epic Train Journeys* (2021), was shortlisted for the National Geographic Traveller Photography Book of the Year, and her new book, *Moonlight Express: Around the World by Night Train*, was published in August 2025.

Felicity Spector has been a television journalist for thirty-five years, starting her career covering the fall of the Communist bloc and the end of the old Soviet Union. She worked as a Moscow bureau producer for ITN between 1990 and 1992 and visited Ukraine when it declared independence. An influential Instagrammer and Substacker, she has built a following of more than 110,000 with a fascinating account that depicts her after-work adventures in restaurants and home baking. She has been spending all her spare time volunteering in Ukraine since summer 2022, helping to support the work of Bake for Ukraine. Her first book is *Bread and War: A Ukrainian Story of Food, War and Hope.* @felicityspector

Sam Williams has lived and worked in Africa and the Middle East. He is executive producer of *Lobito-Bound,* a feature documentary about African railways, mining and geopolitics, and he wrote and produced *Salone Stories,* a podcast series about Sierra Leonean history and culture. Sam is a graduate of the London School of Economics and Oxford University.